MEASUREMENT (CONTINUOUS)	MEASUREMENT (CONTINUOUS)	MEASUREMENT (CONTINUOUS)	MEASUREMENT (CONTINUOUS)	COUNT (DISCRETE)	COUNT (DISCRETE)
Two	Two	Several	One	Two	Several
Unknown	Unknown	Unknown	Known or Unknown	—	—
Large	Small	Large	Large	Large	Large
$(\bar{x}_n - \bar{y}_m)$ $\pm 1.96\,\sigma_{\text{DDSM}}$	$(\bar{x}_n - \bar{y}_m)$ $\pm t_{0.025} \cdot t_{\text{DDSM}}$	—	—	$(\bar{p}_A - \bar{p}_B)$ $\pm 1.96\,\sigma_{\text{DDSP}}$	—
Differences between means		Analysis of variance (ANOVA); F-test	Goodness of fit; chi-squared	Differences between proportions	Contingency tables; chi-squared
$H_0: \mu_A = \mu_B$		$H_0: \mu_A = \mu_B$ $= \mu_C = \cdots$	H_0: Data come from a specified population (e.g., from normal, $\mu = \mu_0,\ \sigma = \sigma_0$)	$H_0: \mu_A = \mu_B$	$H_0: \mu_A = \mu_B$ $= \mu_C = \cdots$
$H_a: \begin{cases} \mu_A \neq \mu_B \\ \text{or} \\ \mu_A > \mu_B \\ \text{or} \\ \mu_A < \mu_B \end{cases}$		H_a: Not all the means are equal	H_a: Data do not come from a specified population	$H_a: \begin{cases} \mu_A \neq \mu_B \\ \text{or} \\ \mu_A > \mu_B \\ \text{or} \\ \mu_A < \mu_B \end{cases}$	H_a: Proportions are not all equal
3-3; 4-4	5-4	6-2; 6-3	4-7	4-5; 3-4	4-6

Quick Symbol Guide

$\bar{x}$	Sample mean (pp. 18, 21, 98–103)
Σ	Summation sign; it means add up the objects that follow (p. 18)
$\displaystyle\sum_{i=1}^{15}$	Add up the 15 objects, numbered 1 to 15, that follow (pp. 18, 21)
$\boxed{\Sigma+}$	A special statistical key on many calculators (p. 18)
$\boxed{\text{2nd}}$	Choose the second function on any two-function key next struck (p. 19)
χ^2, 3 d.f.	Chi-square distribution, or chi-square statistic, with 3 degrees of freedom (pp. 153–155)
$F_{0.01}$ (2, 9)	The 1% critical value of the F-distribution, with 2 numerator degrees of freedom and 9 denominator degrees of freedom (pp. 199, 299)
σ	Standard deviation of the population (p. 53)
σ_{DSM} or $\sigma_{\bar{x}}$	Standard deviation of the distribution of sample means (p. 96)
s	Standard deviation of the sample (p. 53)
$\boxed{x^2}$	The function "squaring" (p. 48)
$|A - B|$	The distance between the numbers A and B (p. 47)
s^2	Sample variance (p. 51)
σ^2 or σ^2_{pop}	Population variance (p. 46)
5!	5 factorial $= 1 \times 2 \times 3 \times 4 \times 5 = 120$ (p. 75)
$\dbinom{5}{3}$	Binomial coefficient (p. 75)
$\cong$	"Is approximately equal to"

Statistics at Your Fingertips

Statistics at Your Fingertips

Mark Finkelstein

University of California, Irvine

Wadsworth Publishing Company

Belmont, California

A division of Wadsworth, Inc.

Statistics Editor: Jim Harrison
Signing Representative: Cindy Champagne
Production Editor: Phyllis Niklas
Designer: Rick Chafian
Copy Editor: Carol Dondrea
Cover: Gary Head

Printed in the United States of America
1 2 3 4 5 6 7 8 9 10   89 88 87 86 85
ISBN 0-534-04023-3

Library of Congress Cataloging in Publication Data

Finkelstein, Mark.
  Statistics at your fingertips.

  Rev. ed. of: Calculate basic statistics/Mark Finkelstein,
George McCarty. c1982.
  Includes index.
  1. Statistics—Data processing.   2. Calculating-
machines.   I. Finkelstein, Mark.   Calculate basic
statistics.   II. Title.
QA276.4.F54   1985   519.5′028′5   84-19655

ISBN 0-534-04023-3

Contents

How to Use this Book

If you've bought this book intending to learn this material on your own, you've taken the right step. I wrote this book with you in mind. (If you're taking a first course in statistics you'll have the guidance of your instructor, so you won't need much guidance from me. But I'm going to give you some anyway.) I wrote this book following many years of experience of teaching "math-phobic" students—students who had the capabilities within themselves of learning this (or more difficult) material, but who somehow thought they didn't. As soon as they saw a symbol they didn't recognize, or an expression they didn't know how to interpret, they threw up their hands and said, "See, I really can't do this, after all!" But some perseverance and, subsequently, some success changed all that: "I really can do this, after all!" Many have been here before you. This material isn't all that difficult, and I've taken special pains to make the explanations clear, and leisurely. It means you have more pages to read through, but you'll find that you will be successful when you do read through them.

THE CALCULATOR YOU NEED TO GET STARTED

In addition to your pencil, you need a calculator. Any model will do—even the cheapest. The only essential functions are addition, subtraction, multiplication, and division. But many brands and models have special features and functions that could be useful to you in your work.

Square-root and *memory* keys are sometimes found on even the calculators sold at drugstores or grocery stores. It will be worth your effort to look for them if you're buying a new machine.

There are *statistical calculators*, such as Texas Instruments's TI-35, which have a special key marked $\boxed{\Sigma}$. This function can cut your button-punching in half, yet the cost of such a calculator is low. Of course, there are much fancier models, whose cost may run to several hundred dollars, and which are programmable. Such machines can be well worth their cost if you frequently do large or complex statistical calculations. You may decide to get one when you finish your study of this book. But a programmable model is definitely not necessary here.

HOW TO USE A COMPUTER AS A CALCULATOR

Let's say that you're in the unusual position of owning a $1000 home computer but lack a $6.95 drugstore calculator. Your computer is almost surely equipped with BASIC (they all are). The single symbol ⎣ ? ⎦ is an instruction to BASIC to print whatever follows. If whatever follows the question mark is an arithmetic expression, the computer will evaluate it and display the result on the screen:

$$?(3 + 4)/10$$

will cause the computer to display 0.7 .

$$?SQR(4 + 6 - 1)$$

will cause the computer to display 3 , since SQR stands for square root. All the arithmetic symbols are as you expect them to be, except that exponentiation is "up arrow": ↑ . So if you want to compute 3^2, type

$$? 3 \uparrow 2$$

and the computer will display 9 . You may make the expressions as complicated as you like. Type

$$? 2 \uparrow (10*7 - 8*8)$$

and the computer will display 64 . Be sure to try these expressions on your computer.

THE LAYOUT OF THE BOOK—EXAMPLES, EXERCISES, PROBLEMS, AND QUIZZES

Everybody knows what examples are. I've made it a point to work out at least one example on each topic that I talk about in the book. I work these examples out for you, down to the last detail. I do this because my experience has been that you never can tell which detail will be the one to "hang the student up," and I want you to be successful. Hopefully, there's enough detail included so that you can see your way through each example, from beginning to end. As for the exercises and problems, the difference between them is that I have completely worked out solutions to the exercises, and these are found at the end of each chapter. The problems have no answers included at all, and so they're suitable for assignment by instructors in a course, or possibly for use on exams. There are usually half a dozen to a dozen problems on each topic. At the end of each chapter I've included a chapter quiz, which is just a little self-test to allow you to check if you've understood the main concepts in the text. Since it's a self-test, I've also included the answers at the end of each chapter. By all means, take advantage of the chapter quizzes. It only takes a couple of minutes to work through each quiz and you'll feel really good when you find you got all the answers right.

THERE IS JUST ONE INSTRUCTION

Do all the exercises. There aren't that many of them—usually just one per topic. You can't possibly learn statistics without *doing* statistics, and doing statistics means working out problems with data. You can't learn statistics by watching me do it any more than you could learn to play the violin by watching me do that. So plan now that you're going to do some work. I've included detailed solutions to each of the exercises (as opposed to the problems) so you can do more than check that you were right or wrong—if you were wrong, you can compare your work with mine, line by line, and *see* where you went wrong. Only by a process like this can you actually learn to do statistics, and learning statistics means learning to *do* statistics. I've done my part, and if you do yours, you'll end up learning statistics. I promise you that *you cannot fail*.

WHAT TO DO WHEN YOU ARE BAFFLED

When you make no sense out of an example or an exercise, the reason is always that you do not understand what every one of the words and symbols means. This is no reason for panic—there is a simple method for handling the situation. Starting at the beginning, double-check each word's definition and each symbol's meaning, one by one. This is easy to do. The first time a new word or symbol is used in this book, it is printed in blacker "boldface" type so you'll notice it. Then it is carefully explained and an example is given of its use. If you have trouble finding this definition, check the Index or use the Quick Word Guide beginning on page 281, the Quick Symbol Guide on page 289, or the Quick Formula Guide on page 290. We guarantee that if you check out all the definitions so you know exactly what is being asked for, *you can work every exercise*.

HOW TO USE THIS BOOK TO REVIEW STATISTICS

If you have already studied statistics, and you want to refresh your understanding of just one topic, look it up in the Index to get started. But this method can be discouraging, so be careful. Suppose your topic is ANOVA, and you find in the Index (or the Contents) that Chapter 6 discusses this technique. You might turn to Chapter 6 and make no sense out of what you read. When this happens, heed the advice above for "what to do when you are baffled."

It will help you in checking back for review to know how topics fit together in this book. Chapters 1, 2, 3, and 4 need to be studied in that order—2 depends on 1, 3 on 2, and 4 on 3. But each of the later chapters, 5, 6, 7, 8, and 9, depends only on the first four. In the example above, of ANOVA, if you aren't prepared to read Chapter 6, do not look at Chapter 5 first. Instead, see if you can do the exercises in Chapter 4. If not, go back to Chapter 3, and so on.

Introduction

YOU CAN'T FAIL AT LEARNING STATISTICS

By buying this book, you have already taken the first step toward mastering statistics. And now, starting with the problem below, you will begin to learn what statistics is all about.

Take the following three numbers

33	45	48

and add them up. Then divide the total by 3. You can work in the space below, if you are going to do it by hand. If you are using a hand calculator, enter the three numbers and add them up; then divide by 3.

Did you get 42 for your answer? Congratulations! You've just observed your first statistic: the *mean*, or average of a *sample*, called simply the *sample mean*. (If you didn't get 42, you either added wrong or divided wrong. Try adding the numbers again and see if you get 126. Then divide 126 by 3 and see if you get 42. If you still don't get the right answer, look at the next page, where I've done the arithmetic for you. Compare your work with mine to see where you went wrong.)

The reason we divide by 3 is that there are 3 numbers in our sample. When we compute the mean, or average, we add up all the numbers in our sample and then divide by the number of numbers. To compute the sample mean of 10 numbers, we add up all 10 numbers and then divide by—you guessed it!—10.

The sample mean is probably the most frequently used statistic. We'll be using it again and again in a variety of settings. But that will come later, after you learn what a statistic is, and what a sample is!

$$\begin{array}{r} 33 \\ 45 \\ +\ 48 \\ \hline 126 \end{array}$$

$$\begin{array}{r} 42 \\ 3\ )\ \overline{126} \\ \underline{12} \\ 6 \\ \underline{6} \end{array}$$

$$\bar{x} = \text{Sample mean} = \frac{33 + 45 + 48}{3}$$

$$= \frac{126}{3} = 42$$

($\bar{x}$ is the symbol we use for the sample mean.)

WHAT THIS BOOK IS ABOUT

Statistics can be divided into two parts: descriptive statistics and inferential statistics. Descriptive statistics deals with collections of numbers, called populations, and the various ways of describing populations. Populations can be described pictorially, by means of bar charts or pie charts, or numerically, by population parameters. Descriptive statistics deals principally with the numerical descriptions—population parameters.

Inferential statistics is the science of making inference, or drawing conclusions, about population parameters when only a portion of the population is available for consideration. Given information about a portion of the population, called the *sample*, we draw conclusions about the population as a whole or about one or more of the population parameters. The development of this science was stymied for a century by the inescapable fact that the conclusions we draw about the population parameters will occasionally be wrong. For years scientists searched for methodology that would allow *some* conclusions to be drawn that would be correct *all* the time. Finally, in the 1920s, it was determined that error was inherent in the nature of inferential statistics; that the error could always be made as small as desired (by paying some price in time or effort), but could never be eliminated. It took several years for this radical notion to gain acceptance in the scientific community, but today it is recognized as the price one has to pay for inferring something about the whole (the population) by studying only a part (the sample).

In this book I will show you how to make inferences about the population mean from the study of two statistics: the sample mean and the sample variance. This one topic contains the main ideas of inferential statistics.

We will then go on to inferences about other parameters from other statistics: chi-square, the *F*-ratio, and *r*. I will also introduce a different method in inferential statistics—that of hypothesis testing. Instead of directly making inferences about a population parameter from a sample statistic, we formulate a hypothesis about a population parameter and then use the sample statistic to *test* the hypothesis (and accept or reject it).

Statistics at
Your Fingertips

Descriptive Statistics: Part 1

Terms we'll be learning about in this chapter

fractile

histogram

mean

median

mode

parameter

percentile

population

statistic

1-1 WHAT STATISTICS IS ALL ABOUT

Statistics is a scientific method for both asking and answering questions about the "real world." These questions are usually of the form, "What is the average annual income of families in this city?" or "Do brand *A* tires last longer than brand *B* tires?" or "Is this drug an effective treatment for this disease?"

While these questions appear to be quite different, they all share something in common: they all ask questions about *numbers*. In the first question, the numbers are the dollars of income of each family. In the second question, the numbers are the miles driven on each kind of tire before it wears out. And in the third question, the numbers are 0s attached to each person who didn't get cured and 1s attached to each person who did. (Don't worry if you can't make sense out of the 0s and 1s at this time.)

Getting back to our first question, we can consider the population we are studying as the set of families in this city. Attached to each object in that set (each family) is a *number*: the number of dollars earned by that family in one year. What we really want to study is the collection of numbers attached to the set of families. Actually, the collection of numbers is not exactly a set, because if two families have the same income, that number would be repeated twice in the collection of numbers. Strictly speaking, the collection of numbers (incomes) we are going to study is a set, together with multiplicities (the two or more copies of a single element, or number).

Our questions about family incomes are really questions about the *numbers* attached to the population of families. We will use the term **population** both for the set of families (the objects) and for the set (with multiplicities) of *numbers* attached to the objects. But usually we use the term *population* to refer to a *set of numbers*.*

Since we will formulate all our statistical questions in terms of populations, we need to develop ways of describing these populations. This area of study is called **descriptive statistics**.

1-2 DESCRIBING POPULATIONS

Some populations are so small we can "comprehend" the entire population, such as the population of coin values (in cents) in my pocket right now. That population is

1, 1, 1, 5, 10, 10, 25, 25

It consists of eight coins, with a total value of 78¢.

But some populations are too large to enumerate comfortably, such as the family incomes in this city. There may be as many as 20,000 families, so the population

---

*To the mathematically sophisticated reader: As you know, a *rule* that attaches a number to each object in some set is really a numerically valued **function** defined on the set. Talking about means, variances, and so on, is really speaking about means and variances of the function. In statistics these functions are called **random variables** (even though there is nothing "random" about them). I won't be using this terminology any more, but if you run into it in another book, you'll know what is being talked about.

would consist of 20,000 elements! And even if we listed all 20,000, we really wouldn't learn much, because all we would have are pages and pages of numbers.

There are many different ways of describing a population without listing all the elements, but basically they fall into two categories: *pictorial descriptions* and *numerical descriptions*. The pictorial descriptions are things like *bar charts* and *pie charts*. The numerical descriptions are called **population parameters**. First we'll deal with the pictorial descriptions. It's a corny old saying, but true: one picture is worth a thousand words.

HISTOGRAMS

Histogram is just another name for *bar chart*, and we'll use the two expressions interchangeably. Just exactly what the heights of the bars mean in Figure 1-1, and what the "cutoff points" for the various classes are, we won't bother specifying at this time. But just from looking at the bar chart in Figure 1-1, we know that the majority of families in that city earn between $15,000 and $30,000 and that few families earn less than $5000. Compare this bar chart (histogram) with the ones shown in Figures 1-2 and 1-3, one for a "rich city" and one for a "poor city." Even

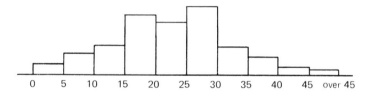

Figure 1-1   Distribution of family incomes in this city, 1982, in thousands of dollars per year.

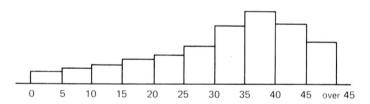

Figure 1-2   "Rich city" incomes in thousands of dollars per year.

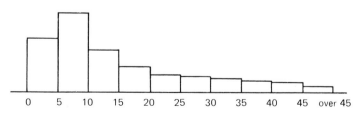

Figure 1-3   "Poor city" incomes in thousands of dollars per year.

without labels, we can tell at a glance which is which. If we were given a list of 20,000 numbers for each of the three cities, however, we would have to spend quite a bit longer figuring out which city was which.

Figures 1-4 and 1-5 show two more examples of histograms; this time the data are taken from the *World Almanac*. Again, we can get quite a bit of information even without looking at the numbers.

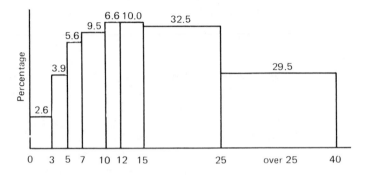

Figure 1-4   White family annual incomes in the United States, 1978, in thousands of dollars. (Data from the *World Almanac*, 1981.)

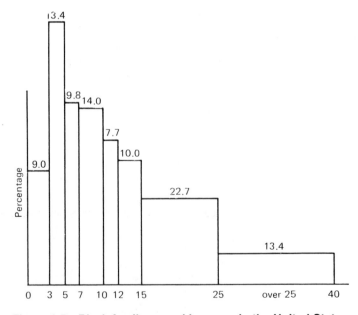

Figure 1-5   Black family annual incomes in the United States, 1978, in thousands of dollars. (Data from the *World Almanac*, 1981.)

PIE CHARTS

Another form of pictorial description is the pie chart. The pie chart in Figure 1-6 shows how the federal budget is divided into different categories of spending. It describes the population of dollars of the total federal budget and shows where each of those dollars goes. The area of each "pie-shaped" piece is proportional to its portion of the budget. You won't encounter any more pie charts in this book, but they are so common in everyday life that they are worth knowing about.

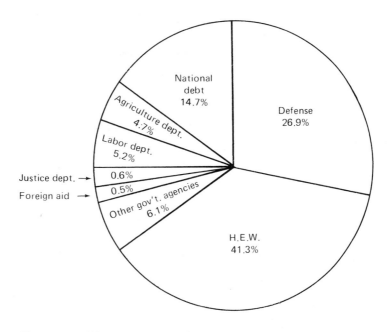

**Figure 1-6   Where the United States Federal Budget Went, 1978.
(Data from _World Almanac_, 1981.)**

PARAMETERS

The bar chart and pie chart are pictorial, or graphical, descriptions of populations. We also have a collection of _numerical_ descriptions of a population. These numerical descriptions are called **parameters**, or population parameters. Examples of population parameters are the **maximum** (the largest element of a population), the **minimum** (the smallest element of a population), the **mean** (the average-sized element of a population), and the **range** (the difference between the maximum and the minimum). There are many other examples of parameters, and we'll get to them in time.

Later in the book you'll learn how to obtain information about a parameter of a population by looking at a sample, which is only a portion of the population. Numbers obtained from looking at the whole population are called parameters, as I've already said. Numbers obtained from looking at only a sample of the population are called **statistics**, or **sample statistics**. And this is what "statistics" means, when you come across a book or course entitled "Statistics."

AN EXAMPLE OF POPULATION PARAMETERS

Table 1-1 gives the annual rainfall in inches in Philadelphia, Pennsylvania, for each of the years 1851 to 1970. This is the complete population. I have listed 120 numbers, and $n = 120$ means the number of numbers is 120, or the size of the population is 120. The notation $n = 120$ just saves you the bother of counting.

Table 1-1   Annual Rainfall in Inches for 120 Years (1851–1970) in Philadelphia, Pennsylvania

35.5	46.3	47.3	30.2	38.2	45.5	51.4	35.4
45.8	45.0	48.4	45.6	34.8	49.8	47.0	29.3
40.7	49.2	55.3	39.2	37.6	41.5	47.4	39.2
40.2	46.0	46.2	39.3	40.3	39.8	39.1	43.1
44.1	56.3	40.2	33.4	31.0	41.6	44.8	32.4
34.0	45.3	47.4	37.2	32.2	51.9	32.3	45.0
48.3	61.2	37.3	42.2	42.0	48.7	39.4	43.2
39.8	51.4	34.5	44.1	49.2	38.1	37.7	39.4
58.1	48.9	36.8	50.6	40.0	37.4	49.1	41.6
44.2	44.1	33.6	34.0	40.9	39.6	46.2	34.0
39.3	32.2	42.0	41.0	38.7	40.9	44.8	40.0
44.5	41.2	51.1	42.6	37.4	52.1	35.0	44.8
51.4	36.8	50.5	35.0	46.9	49.5	47.9	35.5
38.4	39.5	36.9	29.9	45.4	43.3	38.4	43.4
46.4	47.0	33.7	29.3	44.8	45.4	41.2	39.1

$(n = 120)$

Let's construct a histogram for the population of annual rainfall. Count the number of years where the annual rainfall was between 25 and 29.99 inches. Take your time. I can wait. You may find it easier to scan down each column, rather than across the rows.

Did you find three of them? There are two numbers at the bottom of column 4, and one in column 8. Then, 3 years out of 120, or $3/120 = 0.025 = 2.5\%$ of the population lies between 25 and 29.99.

Now we count the number of years with rainfall between 30 and 34.99 inches. And so on. The results are tabulated below:

Inches of Rain	Number of Years	Relative Frequency (Percentage)
25–29.99	3	3/120 = 0.025 = 2.5%
30–34.99	14	14/120 = 0.117 = 11.7%
35–39.99	31	31/120 = 0.258 = 25.8%
40–44.99	31	31/120 = 0.258 = 25.8%
45–49.99	29	29/120 = 0.242 = 24.2%
50–54.99	8	8/120 = 0.067 = 6.7%
55–59.99	3	3/120 = 0.025 = 2.5%
60–64.99	1	1/120 = 0.008 = 0.8%

Here is the first place we encounter "round-off error." While 3/120 = 0.025 *exactly*, 14/120 = 0.117 = 11.7% only *approximately*. Actually, 14/120 = 0.116666 . . . , an infinite repeating decimal that I *rounded off* to 0.117. I chose three places beyond the decimal point as the degree of accuracy that I'm going to work with. The choice is arbitrary. As you work through the book *with your own calculator*, you'll frequently find that your answer and mine will differ in perhaps the third or fourth decimal place. Don't worry about it. The agreement will be close enough.

The information from the list on the previous page can now be displayed by the histogram shown in Figure 1.7.

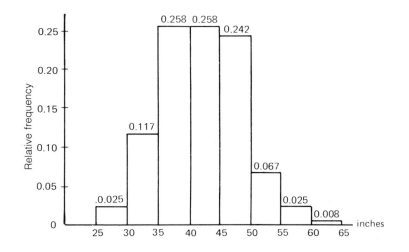

Figure 1-7   Annual rainfall in Philadelphia, 1851–1970.

Now let's note some of the parameters for this population. First find the year with the smallest rainfall. What is it? (It's 29.3, and it occurs in two different years.) So 29.3 is the population *minimum*. The population *maximum* (largest value) is 61.2. The population *range* (maximum minus minimum) is $61.2 - 29.3 = 31.9$. The population *mean* is 42.063 (calculated by adding up all 120 numbers and then dividing by 120). These numerical descriptions of the population—maximum, minimum, range, and mean—are all parameters. Numbers obtained from a *sample*, such as the mean of the sample, are, as we said, called **statistics**.

To see if you understand what you've read so far, take the following quiz. The answers are on page 37.

QUIZ

1. A population is *usually* a collection of

   a. objects
   b. means
   c. statistics
   d. numbers

2. If I have a sample of three numbers and add up the three numbers, then divide the total by 3, I get the

   a. sample mean
   b. population
   c. median

3. The sample mean of 12, 18, 16, and 14 is

   a. 60
   b. 20
   c. 15
   d. I haven't a clue.

4. The population mean, maximum, minimum, and range are examples of

   a. sample statistics
   b. histograms
   c. population parameters

5. Histogram is another name for

   a. number
   b. statistic
   c. bar chart

1-3 HOW TO CONSTRUCT A HISTOGRAM

Here are two examples of populations.

Example 1

Ages of kings and queens of England at coronation, 1600–1981:

37	25	54	32	30	52	39	27	37	54
44	22	58	65	18	60	45	42	41	26

(*n* = 20)   ■

Example 2

Ages of U.S. presidents at inauguration, 1789–1981:

57	61	57	57	58	57	61	54	68	51
49	64	50	48	65	52	56	46	54	49
50	47	55	55	54	42	51	56	55	51
54	51	60	62	43	55	56	61	52	69

(*n* = 40)   ■

I've chosen examples of *small* populations (of numbers) so that they will be easier to deal with. We are usually interested in one or more parameters of the population. In Example 2, the youngest president at inauguration was Theodore

Roosevelt, who succeeded McKinley after the latter's assassination in 1901. He was 42 years old. Therefore, 42 is the minimum of the population in Example 2. In Example 1, the average age at coronation was 40.4, which is, of course, the mean of the population.

When the population is small, we can list all the elements (numbers), as I did in Examples 1 and 2. But when the population is very large (for example, the population of all heights of all males in the United States), we have to resort to descriptions of the population, such as the parameters mean and minimum. Often these parameters are difficult or impossible to obtain, and so we employ statistical inference to *estimate* them. We will deal with statistical inference in Chapter 3.

As you saw earlier, when the population is large, it is often useful to construct a histogram. So now let's take another look at the histogram for annual rainfall in Philadelphia (Figure 1-8). I will now show you how I constructed this histogram, step by step. (Refer to the rainfall data given on page 6.)

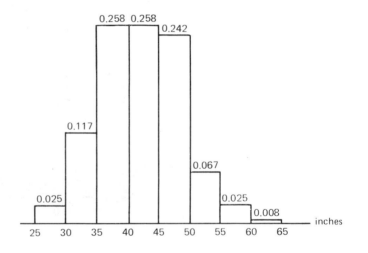

Figure 1-8   Annual rainfall in Philadelphia, 1851–1970.

STEP 1:   DIVIDING THE POPULATION INTO CLASSES

The first thing I had to do was decide how to divide the elements of the population into groups, or **classes**, for purposes of display. Usually, the most sensible system is to group together all the numbers that are relatively close in size. For that reason, I chose my classes as intervals of numbers, such as 25–30, 30–35, and so on. But how did I decide just what size intervals to use, and where to locate them?

I first noted that the numbers in the population varied from a minimum of 29.3

to a maximum of 61.2; therefore, the range was $61.2 - 29.3 = 31.9$. I had to decide how many classes to divide 31.9 into. I didn't want too few classes and I didn't want too many. (What would a histogram look like if it only had two classes?) The generally accepted rule in constructing a histogram is to partition the population into between 6 and 15 classes. If I chose intervals of 5 (inches of rain) each, I would be able to put all the elements of the population into seven such classes, because 5 times 7 is 35, and the range was only 31.9. I could accomplish this with intervals 29–34, 34–39, 39–44, 44–49, 49–54, 54–59, and 59–64. Or, I could accomplish this with intervals 28–33, 33–38, 38–43, 43–48, 48–53, 53–58, and 58–63. As you can see, the exact position of the intervals is somewhat arbitrary. Actually, I prefer to arrange my intervals to fall at the "tens": 30, 40, 50, and so on. And so I chose intervals of 5 inches, starting at 25: 25–30, 30–35, 35–40, 40–45, 45–50, 50–55, 55–60, and 60–65. Note that I have to have eight intervals rather than seven because of where I located them. And note also that I could have constructed the histogram with 4-inch intervals if I had preferred.

With my intervals chosen, I still had one small problem. Where would 30 go? Would I put it in 25–30, or in 30–35? Any consistent rule that decides what to do with boundary elements will work. Generally speaking, we either place all elements that fall on the boundary into the class on the right, or place such elements into the class on the left. You can see how it would have distorted the picture if I had placed the element 30 in the class 30–35 and also placed the element 35 in the class 30–35. I would have given excessive weight to the class 30–35 at the expense of the two neighboring classes.

In constructing my histogram, I decided that all numbers that fell on the boundary of two classes would be placed in the class on the right. However, this meant taking the right-hand endpoint of each class to more decimal places of accuracy than the data, as follows:

Class: Between 25 and 29.99
           Between 30 and 34.99
               etc.

Once I did this, I no longer had any conflict as to where the number 30 would go. It could not belong to the class 25–29.99. It could go only into the class 30–34.99. I thus arrived at the following eight classes:

Between 25 and 29.99
   "   30   "   34.99
   "   35   "   39.99
   "   40   "   44.99
   "   45   "   49.99
   "   50   "   54.99
   "   55   "   59.99
   "   60   "   64.99

STEP 2:   MAKING A TALLY SHEET

Once I had my classes, my next step was to "tally," or make a mark for each element next to the class in which it belongs. Since the first element in the population of rainfall was 35.5, I made a mark, or tally, on the tally sheet as follows:

25–29.99
30–34.99
35–39.99
40–44.99
45–49.99
50–54.99
55–59.99
60–64.99

The second number in the population (reading down) was 45.8. I made another mark, this time next to the interval 45–49.99, and my tally sheet became

25–29.99
30–34.99
35–39.99
40–44.99
45–49.99
50–54.99
55–59.99
60–64.99

I continued in the same way until I had tallied all 120 numbers on my tally sheet. Notice that I grouped tallies in groups of five by marking the fifth tally *across* the preceding four tallies:

卅

Clearly, 13 tallies are easier to count this way

卅 卅 ///

than this way

/////////////

Here is my completed tally sheet:

25–29.99	///
30–34.99	ᵗᴴᴸ ᵗᴴᴸ ////
35–39.99	ᵗᴴᴸ ᵗᴴᴸ ᵗᴴᴸ ᵗᴴᴸ ᵗᴴᴸ ᵗᴴᴸ /
40–44.99	ᵗᴴᴸ ᵗᴴᴸ ᵗᴴᴸ ᵗᴴᴸ ᵗᴴᴸ ᵗᴴᴸ /
45–49.99	ᵗᴴᴸ ᵗᴴᴸ ᵗᴴᴸ ᵗᴴᴸ ᵗᴴᴸ ////
50–54.99	ᵗᴴᴸ ///
55–59.99	///
60–64.99	/

STEP 3: COMPUTING THE RELATIVE FREQUENCIES

I then converted the tallies to counts by counting how many there were in each class. Then I computed the relative frequencies and percentages of each class by dividing the total count in each class by 120, the total number of elements in the population. Then I multiplied the relative frequency by 100 to obtain a percentage:

Class	Tallies	Counts
25–29.99	///	3
30–34.99	ᵗᴴᴸ ᵗᴴᴸ ////	14
35–39.99	ᵗᴴᴸ ᵗᴴᴸ ᵗᴴᴸ ᵗᴴᴸ ᵗᴴᴸ ᵗᴴᴸ /	31
40–44.99	ᵗᴴᴸ ᵗᴴᴸ ᵗᴴᴸ ᵗᴴᴸ ᵗᴴᴸ ᵗᴴᴸ /	31
45–49.99	ᵗᴴᴸ ᵗᴴᴸ ᵗᴴᴸ ᵗᴴᴸ ᵗᴴᴸ ////	29
50–54.99	ᵗᴴᴸ ///	8
55–59.99	///	3
60–64.99	/	1

Relative Frequencies	Percentage
3/120 = 0.025	2.5%
14/120 = 0.117	11.7%
31/120 = 0.258	25.8%
31/120 = 0.258	25.8%
29/120 = 0.242	24.2%
8/120 = 0.067	6.7%
3/120 = 0.025	2.5%
1/120 = 0.008	0.8%
	100.0%

Occasionally when we make this computation, the totals do not come out exactly to 100.0%. This is because the decimals we compute run to more places than the accuracy we carry in our computation. (For example, $14/120 = 0.11666667$, to eight decimal places, but I rounded it off to 0.117.) It is not unusual for the total to come to a little more or less than 100% due to round-off error. This will not be the case when the number of elements in the population is 5, 10, 20, 25, 50, or 100; but it will be the case when the number of elements is a number like 6, 14, or 21. Don't worry if you get 0.996 or 1.002 for your total. But if you get 0.983 or 1.025, you'd better check your arithmetic.

STEP 4: DRAWING THE BAR CHART

I now had all the information I needed. The first thing I drew was the horizontal axis, labeling the axis with the beginning value of each class. I no longer needed the numbers 29.99, 34.99, and so on. Here is how the axis looked:

25   30   35   40   45   50   55   60   65

The last step was to draw in the bars. It is customary to arrange the bars so that the area of each bar is proportional to the relative frequency of the corresponding class. Since all the classes in this example have the same width (interval), making the area of each bar proportional to the relative frequency of the class was accomplished by making the height of each bar proportional to the relative frequency. Figure 1-9 shows the finished histogram.

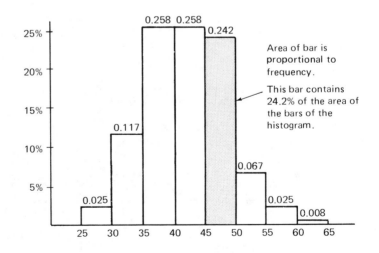

Figure 1-9

How large we make the vertical scale, and so how high we make the bars, is arbitrary. We could make all the bars twice as high, or half as high, and the histogram would convey the same information. We simply use whatever scale is most convenient.

It is possible to construct a histogram in which the intervals are not equal. (Figures 1-4 and 1-5 are examples of such histograms.) But we must remember to make the areas of each bar proportional to the relative frequency of the class; thus, we must arrange the heights accordingly. If, for example, we were to combine classes 30−35 and 35−40 of Figure 1-9 into one class, 30−40, its count would be 45, and so its relative frequency would be $45/120 = 0.375 = 37.5\%$. But because the interval 30−40 is *twice* as wide as the others, to make the *area* of that bar proportional to the relative frequency, we would have to make the bar *one-half* as high as we would if the intervals were of equal width, and the histogram would look like the one in Figure 1-10.

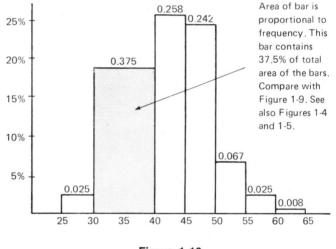

Area of bar is proportional to frequency. This bar contains 37.5% of total area of the bars. Compare with Figure 1-9. See also Figures 1-4 and 1-5.

Figure 1-10

EXERCISE 1   Construct a histogram for the population of Example 2, the ages of U.S. presidents at inauguration, 1789−1981. Refer to the data on page 9, and use intervals 40−45, 45−50, 50−55, 55−60, 60−65, and 65−70. Decide what to do with the numbers that fall on the boundaries of the classes. (Do the same thing we did with our worked-out example.) Use the worksheet provided on page 16. The answers to the exercises appear at the end of each chapter.

Worksheet for Exercise 1

Class	Tally	Count	Proportion	Percentage (Relative Frequency)
(40 – 44.99) ll		2	.05	5%
(45 – 49.99) ┼┼┼┼		5	.125	12.5%
(50 – 54.99) ┼┼┼┼ ┼┼┼┼ l		11	.275	27.5%
(55 – 59.66) ┼┼┼┼ ┼┼┼┼ l		11	.275	27.5%
(60 – 64.99) ┼┼┼┼ l		6	.15	15%
(65 – 69.4) lll		3	.075	7.5%

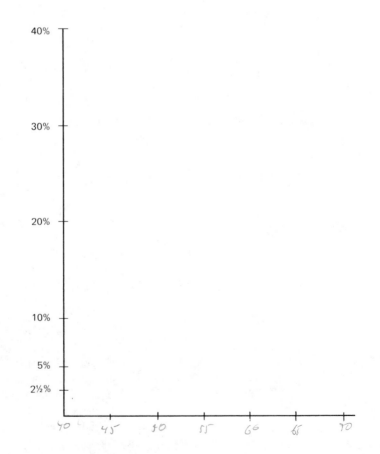

1-4 MEASURES OF LOCATION: THE MEAN

Most populations are too large for us to list all the elements. For example, we might be interested in the population of all heights (in inches) of all U.S. males over the age of 21. We have seen that a histogram is a useful pictorial device for summarizing such information. But statistics is a science of numbers, and we need numerical descriptions of populations as well as pictorial ones.

Figure 1-11 shows histograms of two imaginary populations, population *A* and population *B*. How would you describe the difference between these two populations?

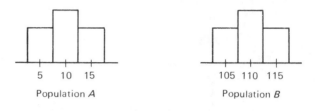

Figure 1-11

The numbers in population *A* are grouped about 10, while the numbers in population *B* are grouped about 110. The two populations differ in the *sizes* of their numbers, that is, in their *location* along the number line (see Figure 1-12). There are several numerical descriptions of a population that would reflect the fact that population *A* is located about 10 while population *B* is located about 110. The simplest and most commonly used such measure is the *mean*, or arithmetic mean, of the population.

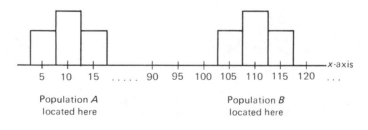

Figure 1-12

The **mean** is the arithmetic average of all the numbers in a population. We obtain the mean by adding up all the numbers in a population and then dividing by the number of numbers. The symbol for the mean is $\bar{x}$.

Example 3

The mean of the numbers $\boxed{33 \qquad 45 \qquad 48}$ is

$$\frac{33 + 45 + 48}{3} = \frac{126}{3} = 42 \qquad \blacksquare$$

I might as well get you familiar with mathematical notations right now. If we let x_1, x_2, x_3 stand for the three numbers (33, 45, 48), then the symbol for adding up the three of them is

$$\sum_{i=1}^{3} x_i.$$

The Σ stands for "sum" and is the Greek letter sigma. Note that i is "varying" from 1 (the lower limit of $\sum_{i=1}$) to 3 (the upper limit of $\overset{3}{\Sigma}$), so $i = 1$, 2, and 3 in turn. Thus, we could write

$$\bar{x} = \frac{\sum_{i=1}^{3} x_i}{3},$$

which we know now means the same as

$$\bar{x} = \frac{x_1 + x_2 + x_3}{3}.$$

Frequently we'll drop the upper and lower limits of the Σ and just write

$$\bar{x} = \frac{\sum x_i}{3}$$

when it's clear how many there are to add up, and which ones. The symbol Σ is the same symbol that appears on statistical calculators.

IF YOUR CALCULATOR HAS A $\boxed{\Sigma+}$ KEY

If you have a calculator such as the TI-35, or other calculator designed to handle basic statistical calculations, note that it has a special key for entering statistical data and computing means and standard deviations. This key is usually denoted $\boxed{\Sigma+}$. If your calculator does not have this function, and is a simple four- or six-function calculator, skip this section and read below where we talk about four- and six-function calculators.

If your calculator is on, you must first clear the statistical registers before beginning the calculation. Each calculator has a different sequence to accomplish this. On some calculators it's a key marked ⎡CM⎤ ("clear memory"); on others it's a key marked ⎡CLEAR⎤ (as opposed to ⎡CLx⎤). On the TI-35, it's a key marked ⎡CSR⎤ (for "clear statistical registers"), which is a second function on a two-function key (marked ⎡x!/CSR⎤). Because it's the *second* function I want, I first press the key marked ⎡2nd⎤ and then the key marked ⎡CSR⎤ to obtain cleared statistical registers. So, when I say "clear statistical registers," I'll use the keystroke sequence ⎡2nd⎤ , ⎡CSR⎤ ; but you should do whatever is required on *your* calculator to accomplish this. (See your calculator manual for instructions.)

If your calculator is already on, press ⎡2nd⎤ , ⎡CSR⎤ , to be sure to clear the memory. Now enter the first number, 33, by keying ⎡3⎤ followed by ⎡3⎤ . The calculator should be displaying 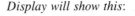 *33* (see Figure 1-13). Now press ⎡Σ+⎤ . The calculator should now be displaying *1* , as you have entered one number. (If your calculator is displaying any number other than

Press this key on your TI-35:	*Display will show this:*
2nd	*?*
CSR	*?*
3	*3*
3	*33*
Σ+	*1*
4	*4*
5	*45*
Σ+	*2*
4	*4*
8	*48*
Σ+	*3*
2nd	*3*
STO/x̄	*42*

Figure 1-13

1   at this point, turn your calculator off, turn it back on, and start this section again.) Now key in the second number, 45 (4 , 5) followed by Σ+ . The calculator now displays   2 , as you have just entered the second number. Now enter 48 (4 , 8), followed by Σ+ . The calculator displays   3 , as you have just entered the third number. Now press 2nd followed by x̄ (the symbol for the *mean*). (The reason we press 2nd before we press x̄ is that the calculator uses the same key for two different functions. The x̄ key is also the STO key on the TI-35. If you press the key normally, you STOre the number you are displaying. The function we wish to perform is the calculation of x̄, or the *mean*, which is the *second function* of the key marked STO/x̄ . We tell the calculator that we wish to perform the second function by pressing the 2nd key before we press the STO/x̄ key.)

Is your calculator displaying   42 ? Congratulations! You have just used the statistical features of your calculator to determine the mean of the population that you worked with in Example 3.

IF YOU HAVE A FOUR-FUNCTION OR SIX-FUNCTION CALCULATOR (NO Σ+ KEY)

Turn your calculator on. If it is already on, press CLR to clear the accumulator and display (see Figure 1-14). Now press 3 followed by 3 . The calculator should be displaying   33 , the first number. (If it isn't, turn the calculator off, then on again, and start again.) Now press the + key, as you are going to *add* up numbers. The calculator should still be displaying   33 . Then key in 45 (4 , 5), and press the + key again. The calculator should now be displaying   78 , the sum of the first two numbers. Then key in 48 (4 , 8), followed by = , which should bring your total to 126. The calculator should now be displaying   126 . Now we wish to divide this total by 3, so press ÷ , which indicates that we wish to divide the number that is displayed by the number that will follow, in our case, 3. So now key in 3 , followed by = . The display should be   42 , which is the mean of our population of three numbers.

Example 4

The mean of the population of Example 1, the ages of kings and queens of England at coronation, 1600–1981, is

$$\frac{\begin{array}{l} 37 + 25 + 54 + 32 + 30 + 52 + 39 + \\ 27 + 37 + 54 + 44 + 22 + 58 + 65 + \\ 18 + 60 + 45 + 42 + 41 + 26 \end{array}}{20} = \frac{808}{20} = 40.4 \text{ years}$$

The number of numbers

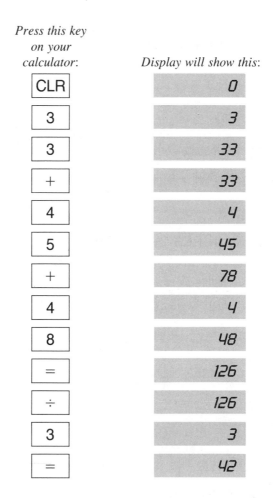

Figure 1-14

Thus, the average, or mean, age is 40.4 years. Symbolically, $\bar{x} = 40.4$. If x_1, $x_2, x_3, \ldots, x_{20}$ are symbols for each of the 20 ages, then the mean is denoted by

$$\bar{x} = \frac{x_1 + x_2 + x_3 + \cdots + x_{19} + x_{20}}{20} = \frac{\displaystyle\sum_{i=1}^{20} x_i}{20}.$$

Use your calculator to check this computation. Of course, I divided by 20 because I added up 20 numbers; if I added up 30 numbers, I'd divide by 30:

$$\bar{x} = \frac{\displaystyle\sum_{i=1}^{30} x_i}{30} = \frac{\sum x_i}{30}.$$

(Recall that I said we'd drop the limits on Σ when they were obvious.) Or, if I added up n numbers, I'd divide by n:

$$\bar{x} = \frac{\sum\limits_{i=1}^{n} x_i}{n} = \frac{\sum x_i}{n}.$$

EXERCISE 2

Calculate the mean age of the U.S. presidents at inauguration, 1789–1981, the population of Example 2, page 9. Use the features of your calculator.

$\bar{x} = $ _____

1-5 OTHER MEASURES OF LOCATION

While the mean has many good features as a measure of location for a population, it suffers from one drawback: it is affected too much by extreme values. To illustrate, suppose we examine the 10 sales receipts for a small store one afternoon:

9.83	10.05	4.95	8.31	0.95
6.42	8.41	12.09	11.86	15.91

$$(n = 10)$$

We compute the mean, $\bar{x} = $ _____. (Use your calculator to obtain this, and fill in the blank.)

Did you get 8.878 as your answer? Now suppose that one of the values, 8.31, had been incorrectly recorded as 831.00. How would this affect the mean?

IF YOU HAVE A FOUR-FUNCTION OR SIX-FUNCTION CALCULATOR

Compute the mean by adding the 10 numbers above, but change 8.31 to 831. You can either key in all 10 numbers again, or if you have the original sum available, subtract 8.31 and add 831; then divide by 10.

IF YOU HAVE A $\boxed{\Sigma+}$ CALCULATOR

If your calculator is still on, displaying $\bar{x} = $ ▓▓▓▓ *8.878* , you can recompute the mean without reentering all 10 numbers. To accomplish this on the TI-35 there is a "companion" key to the $\boxed{\Sigma+}$ key. It is the $\boxed{\Sigma-}$ key. This key is used to delete incorrectly keyed entries. To delete 8.31, key in 8.31, then press $\boxed{2nd}$ followed by $\boxed{\Sigma-}$ (see Figure 1-15). The 8.31 is removed from the entries,

Key:	Display:
	8.878
8	8
.	8
3	8.3
1	8.31
2nd	8.31
$\Sigma-$	9
8	8
3	83
1	831
$\Sigma+$	10
2nd	10
$\bar{x}$	91.147

Figure 1-15

and the calculator should be displaying 9 , the number of entries remaining. Now enter 831 and press $\boxed{\Sigma+}$. If your calculator is displaying 10 , you have successfully deleted 8.31 and replaced it with 831. If you now press $\boxed{2nd}$ followed by $\boxed{\bar{x}}$, the calculator should be displaying the new mean, 91.147 .

Now observe what we did when we changed just one entry of our 10 numbers:

			831	
9.83	10.05	4.95	8.31	0.95
6.42	8.41	12.09	11.86	15.91

The mean changed from 8.878 to 91.147. So, if we are using the mean as a measure of the average *size* of a sale at the store on a particular afternoon, one single changed entry can multiply the mean (in this case) 10 times. This is why we say the mean is very sensitive to extreme values.

A measure of location that is not sensitive to extreme values is the *median*. The **median** is the center, or middle, value when the numbers in a population are arranged in increasing or decreasing order. If there are two middle values (as would be the case if there were an *even* number of numbers in the population), then the median is defined as the average of the two middle numbers.

Let's compute the *median* of our 10 register receipts, and recompute it after we change 8.31 to 831.00:

	Original Receipts Ranked	Changed Receipts Ranked	
	0.95	0.95	
	4.95	4.95	
	6.42	6.42	
Median	8.31	8.41	*Median*
$\dfrac{8.41 + 9.83}{2}$	8.41	9.83	$\dfrac{9.83 + 10.05}{2}$
	9.83	10.05	
$= 9.12$	10.05	11.86	$= 9.94$
	11.86	12.09	
	12.09	15.91	
	15.91	831.00	

So the original median was 9.12, which increased to 9.94 when we changed the one entry 8.31 to 831. While the median did change, it did not change as drastically as did the mean, which increased by a factor of 10 or more. This is why we say that the median is less sensitive to extreme values than the mean. The median is often used to estimate the mean in situations where it is known, or suspected, that extreme values can occur. (Another method for estimating the mean is to discard the largest and smallest values in a sample. This is what is done in international diving competitions, where the highest and lowest judge's scores are discarded and the mean of the remaining scores computed.)

The median is the number such that 50% of the population is smaller and 50% of the population is larger. For that reason, the median is also called the **50th percentile**. Can you guess what the 75th percentile is? It is the number such that 75% of the population is smaller and 25% is larger. Similarly, the 99th percentile is the number such that 99% of the population is smaller and 1% is larger. Notice also that the 99th percentile need not be a number in the population, just as long as it is a number such that 99% of the population is smaller and 1% is larger.

Also, the *first quartile* (Q_1) is the number such that 1/4 of the population is smaller and 3/4 is larger; the *second decile* (D_2) is the number such that 2/10 of the population is smaller and 8/10 of the population is larger; the *fourth quintile* (QN_4)

is the number such that 4/5 of the population is smaller and 1/5 larger. *Percentile*, *quartile*, *quintile*, and *decile* are all instances of **fractiles** of a population.

It is sometimes difficult to find exactly which number represents the median or other fractile. As I said, it needn't be an element of the population. For an even number of numbers in the population, as already noted, the median is defined as the average of the two "middle" values. But for a small population of numbers, like the 10 register receipts listed, talking about the 75th percentile is not very meaningful. Percentiles are useful only when we are talking about very large populations (like all College Board scores in the United States, a population that numbers in the millions). For small populations, the simplest way to define the various fractiles of the population is by a histogram.

When we construct our histogram, we arrange it so that the *areas* of the bars are proportional to the frequencies of the classes. Then we can define the median as the (unique) point where 50% of the area of the histogram is to the right and 50% is to the left (see Figure 1-16a). The 75th percentile is, similarly, the unique point on the *x*-axis where 75% of the area of the bars is to the left and 25% of the area is to the right (see Figure 1-16b). I will show you how to compute the median and other fractiles also by using ogives, which we will be discussing below.

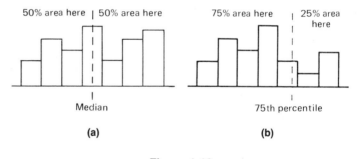

Figure 1-16

You may wonder whether our two definitions of median—one involving the numbers in the population directly, the other involving the histogram of the population—are the same. Actually, they aren't. To construct a histogram requires making choices about how to group numbers into classes, and once we do that, we "throw away" some information about the population. The number 2.5 and the number 2.9 both are in the class 2–2.99. They are two different numbers, but they each just count one tally when we construct the histogram. In practice, we don't construct histograms for very small populations, like the 10 register receipts. And for large populations, like all U.S. College Board scores, the two definitions of median produce essentially the same result.

Another method for indicating the location of a population is to report the 25th and 75th percentiles (which is the same as the first and third quartiles). This is very similar to the method of discarding extreme values. By reporting these two numbers, we announce where the central 50% of the population is located.

There's another device that statisticians use to describe populations, one that we can use to compute the various fractiles of the population at the same time. It is called the **ogive**, or cumulative distribution of the population. Roughly speaking, the ogive gives the percentage of the population that is *less* than a given value.

We can start with the histogram of a population, say the rainfall histogram given in Figure 1-8 or Figure 1-9, and convert it into an ogive of the population, as follows: the histogram is a *function* that gives, for each class or interval, the relative frequency (percentage of the population) that falls *within* that interval. The *ogive* of the population is the function that gives, for each class or interval, the relative frequency that is *less than* the right-hand endpoint of the class or interval. For this reason we call the values of the ogive cumulative frequencies (they are the sum of all the values that came before).

Starting with the relative frequencies of our rainfall data as displayed in the histogram of Figure 1-9, we compute the following cumulative frequencies:

Class	Relative Frequency	Cumulative Frequency	
25–29.99	0.025	0.025	Less than 30
30–34.99	0.117	0.142	Less than 35
35–39.99	0.258	0.400	Less than 40
40–44.99	0.258	0.658	Less than 45
45–49.99	0.242	0.900	Less than 50
50–54.99	0.067	0.967	Less than 55
55–59.99	0.025	0.992	Less than 60
60–64.99	0.008	1.000	Less than 65

The cumulative frequency for less than 30 inches is the percentage or proportion of the population that is less than 30 inches, which in this case is all of the class 25–29.99, totaling 2.5% = 0.025. The cumulative frequency for less than 35 is the sum of the percentages in the class 25–29.99 and the class 30–34.99, or 0.025 + 0.117 = 0.142. That is, 14.2% = 0.142 of the population is rainfall of less than 35 inches of rain. We continue in the same way: a rainfall of less than 40 inches of rain is in class 25–29.99 or 30–34.99 or 35–39.99. These three classes have relative frequencies of 0.025, 0.117, and 0.258, respectively. So the percentage of the population that is less than 40 = 0.025 + 0.117 + 0.258 = 0.400. Continuing in the same way, we compute the cumulative frequencies for each of the classes. Notice that since the largest rainfall was 61.2 inches, when we compute the proportion of the population that is less than 65 inches, we get 100% = 1.000 relative frequency.

Next we draw an axis that looks like this:

25   30   35   40   45   50   55   60   65

just as we did when we were drawing our histogram. Then we place a point above each of these values at the height that indicates the cumulative frequency (noting that the cumulative frequency of the value 25 inches is 0, since the percentage of the years that had less than 25 inches of rainfall is 0.000%). We get a picture that looks like Figure 1-17.

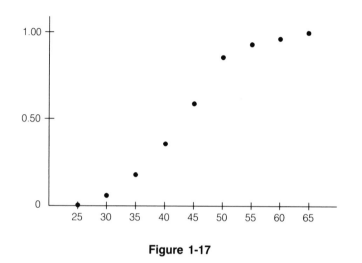

Figure 1-17

Now we draw straight lines to connect the points (a process called *linear interpolation*), and we have our completed ogive, Figure 1-18.

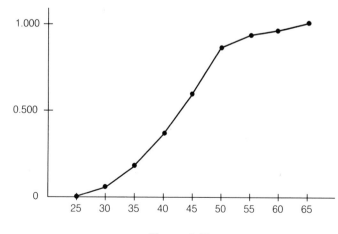

Figure 1-18

HOW TO USE THE OGIVE TO COMPUTE THE MEDIAN AND OTHER FRACTILES

Now that we've constructed the ogive, it's a simple matter to calculate the median. Recall that we said the median was the value for which 50% of the population was greater and 50% was less. Now look at our ogive, and remember what the ogive tells us (the percentage of the population that is *less than* a given value). To find the median, we have only to look for that value on the horizontal axis, where the value of the cumulative frequency is 0.500 (50%). That's easy to find. See Figure 1-19. All we have to do is start at 0.500 on the vertical scale and backtrack to find the value on the horizontal axis that corresponds to it. Then we've found the value where 50% of the population is smaller and 50% is larger. This same idea works for any fractile of the population.

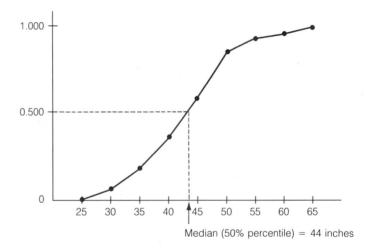

Figure 1-19   50% of the population lies to the left of 44, and 50% lies to the right.

On the ogive shown in Figure 1-20 I've displayed the 30th and 75th percentiles for you, following the same method.

It is also possible to give a formula for computing these values more precisely, but this is unnecessarily complicated—usually the values we obtain "by eye" are good enough. Also, since our ogive was constructed with "grouped data," our fractile computations will only be approximately correct at best.

EXERCISE 3   Compute the median age of kings and queens of England at coronation, Example 1,

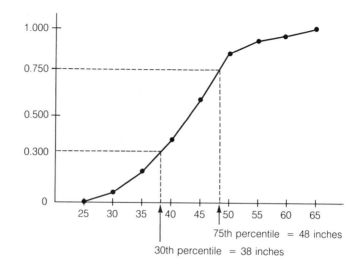

Figure 1-20 These values were obtained by "reading" the ogive "by eye."

and the median age of U.S. presidents at inauguration, Example 2. The data are on page 9.

Median age of kings and queens = _____.

Median age of presidents = _____.

Still another measure of location is the mode or modal class. The **mode** of a population is the number that occurs most often. The **modal class** is the class in a histogram that occurs most often. In Example 1, the ages of kings and queens, there are two modes, 37 and 54, which occur twice each. When we examine the histogram we constructed for the rainfall example (Figure 1-7), we see that there are two modal classes, 35–40 and 40–45. When there are two modes for a population, we call the population **bimodal**. When a population is bimodal, it often suggests that there are two subpopulations involved. This is not the case in the ages of kings and queens; but if we consider the frequently used example of a population of heights of individuals, we will find that this population is often bimodal. Why would we expect such a population to be bimodal? What would be the two subpopulations in this case?

(*Answer:* Men's heights and women's heights.)

EXERCISE 4   For Example 1 and Example 2, page 9, what is the first quartile, fourth quintile, seventh decile?

_____   _____   _____

EXERCISE 5

What is the mode of the population of Example 2, the ages of U.S. presidents at inauguration? The data are on page 9.

_____

EXERCISE 6

Referring to the histogram you constructed in Exercise 1, page 16, what is the modal class? What is the 50th percentile (median)?

Modal class _____

50th percentile _____

EXERCISE 7

Construct an ogive for the data in Example 2, ages of U.S. presidents. The data are on page 9. Use as intervals the same intervals you used in the worksheet in Exercise 1, page 16.

EXERCISE 8

Using the ogive you constructed in Exercise 7, find the median, the 25th percentile, and the 80th percentile for the ages of U.S. presidents.

PROBLEM SET A

1. Compute the mean of the following population of 50 corporation presidents' ages:

58	59	63	61	62	47	56	62	63	51
69	58	67	68	59	67	60	60	54	63
54	67	66	57	65	57	55	58	65	51
59	70	63	59	74	71	58	51	43	44
61	59	61	70	61	63	54	60	60	71

2. Compute the mean of the following population of 26 daily expenses:

44.98	66.56	49.28	68.03	54.95	58.49	54.68
51.37	39.06	50.05	47.57	61.20	72.60	77.11
41.24	47.97	71.35	42.91	65.03	57.54	43.54
69.17	45.39	38.89	59.97	36.43		

3. Compute the mean of the following population of 45 male models' heights, in inches:

71	72	70	71	72	72	70	70	71	70
73	72	70	72	70	74	72	72	73	71
71	72	71	73	72	71	72	71	74	71
71	71	74	73	72	71	72	70	71	71
73	70	70	72	72					

4. Compute the mean of the following population of 15 students' attention spans, in minutes:

10.9	10.1	9.2	13.6	9.9	9.7	13.8	13.0
5.0	6.6	8.6	14.0	8.5	9.9	10.1	

5. Compute the mean of the following population of 30 automobile gasoline mileages:

21.9	38.6	34.7	31.2	17.2	18.6	28.4	30.3
35.7	21.9	23.1	38.8	38.0	37.3	16.2	22.7
28.3	22.4	24.0	15.6	31.5	22.7	24.4	16.9
26.6	21.8	30.6	19.6	35.8	23.0		

6. Compute the mean of the following population of 35 television repair bills:

60.33	55.82	55.99	165.58	96.48
154.74	137.06	108.82	67.42	79.94
87.44	159.72	88.50	175.32	165.41
153.53	104.00	178.08	49.35	65.01
119.59	121.20	130.35	130.98	154.85
155.94	124.60	52.77	75.76	94.34
126.01	158.76	107.22	111.72	98.59

7. Compute the mean of the following population of 15 rats' weight gains, in grams:

8.8	28.3	9.6	9.6	42.0	43.3	5.3	19.6
34.2	13.2	4.2	45.9	5.9	12.6	10.7	

8. Compute the mean of the following population of 32 executive salaries:

128,000	145,000	78,000	136,000	60,000	72,000
91,000	217,000	46,000	191,000	85,000	127,000
139,000	134,000	182,000	104,000	173,000	177,000
197,000	213,000	149,000	112,000	133,000	145,000
101,000	155,000	167,000	143,000	213,000	88,000
149,000	173,000				

9. Compute the mean of the following population of 60 pocketsful of change, in cents:

7	18	0	89	4	24	6	21	21	15
19	45	55	71	32	29	54	62	33	38
11	79	69	51	86	76	60	63	28	72
0	32	57	18	85	27	63	48	58	37
86	25	15	74	5	21	62	26	68	1
87	81	84	71	1	78	83	30	8	69

10. Compute the mean of the following population of 30 flying times (in minutes) to cross the United States:

334	317	290	289	285	264	294	280	324	301
306	316	286	328	358	276	358	308	325	354
289	281	253	268	328	270	268	318	289	332

11. Compute the mean of the following population of 33 measurements of galvanic skin response, in volts:

0.0213	0.0182	0.0201	0.0188	0.0204
0.0181	0.0190	0.0198	0.0214	0.0204
0.0208	0.0206	0.0210	0.0206	0.0200
0.0215	0.0206	0.0199	0.0206	0.0210
0.0190	0.0192	0.0202	0.0218	0.0181
0.0217	0.0180	0.0180	0.0182	0.0201
0.0197	0.0190	0.0214		

PROBLEM SET B

1. Compute the median of the following population of 50 corporation presidents' ages:

58	59	63	61	62	47	56	62	63	51
69	58	67	68	59	67	60	60	54	63
54	67	66	57	65	57	55	58	65	51
59	70	63	59	74	71	58	51	43	44
61	59	61	70	61	63	54	60	60	71

2. Compute the median of the following population of 26 daily expenses:

44.98	66.56	49.28	68.03	54.95	58.49	54.68
51.37	39.06	50.05	47.57	61.20	72.60	77.11
41.24	47.97	71.35	42.91	65.03	57.54	43.54
69.17	45.39	38.89	59.97	36.43		

3. Compute the median of the following population of 45 male models' heights, in inches:

71	72	70	71	72	72	70	70	71	70
73	72	70	72	70	74	72	72	73	71
71	72	71	73	72	71	72	71	74	71
71	71	74	73	72	71	72	70	71	71
73	70	70	72	72					

4. Compute the median of the following population of 15 students' attention spans, in minutes:

10.9	10.1	9.2	13.6	9.9	9.7	13.8	13.0
5.0	6.6	8.6	14.0	8.5	9.9	10.1	

5. Compute the median of the following population of 30 automobile gasoline mileages:

21.9	38.6	34.7	31.2	17.2	18.6	28.4	30.3
35.7	21.9	23.1	38.8	38.0	37.3	16.2	22.7
28.3	22.4	24.0	15.6	31.5	22.7	24.4	16.9
26.6	21.8	30.6	19.6	35.8	23.0		

6. Compute the median of the following population of 35 television repair bills:

60.33	55.82	55.99	165.58	96.48
154.74	137.06	108.82	67.42	79.94
87.44	159.72	88.50	175.32	165.41
153.53	104.00	178.08	49.35	65.01
119.59	121.20	130.35	130.98	154.85
155.94	124.60	52.77	75.76	94.34
126.01	158.76	107.22	111.72	98.59

7. Compute the median of the following population of 15 rats' weight gains, in grams:

8.8	28.3	9.6	9.6	42.0	43.3	5.3	19.6
34.2	13.2	4.2	45.9	5.9	12.6	10.7	

8. Compute the median of the following population of 32 executive salaries:

128,000	145,000	78,000	136,000	60,000	72,000
91,000	217,000	46,000	191,000	85,000	127,000
139,000	134,000	182,000	104,000	173,000	177,000
197,000	213,000	149,000	112,000	133,000	145,000
101,000	155,000	167,000	143,000	213,000	88,000
149,000	173,000				

9. Compute the median of the following population of 60 pocketsful of change, in cents:

7	18	0	89	4	24	6	21	21	15
19	45	55	71	32	29	54	62	33	38
11	79	69	51	86	76	60	63	28	72
0	32	57	18	85	27	63	48	58	37
86	25	15	74	5	21	62	26	68	1
87	81	84	71	1	78	83	30	8	69

10. Compute the median of the following population of 30 flying times (in minutes) to cross the United States:

334	317	290	289	285	264	294	280	324	301
306	316	286	328	358	276	358	308	325	354
289	281	253	268	328	270	268	318	289	332

11. Compute the median of the following population of 33 measurements of galvanic skin response, in volts:

0.0213	0.0182	0.0201	0.0188	0.0204
0.0181	0.0190	0.0198	0.0214	0.0204
0.0208	0.0206	0.0210	0.0206	0.0200
0.0215	0.0206	0.0199	0.0206	0.0210
0.0190	0.0192	0.0202	0.0218	0.0181
0.0217	0.0180	0.0180	0.0182	0.0201
0.0197	0.0190	0.0214		

PROBLEM SET C

1. Compute the third quartile, second quintile, and the ninth decile of the following population of 50 corporation presidents' ages:

58	59	63	61	62	47	56	62	63	51
69	58	67	68	59	67	60	60	54	63
54	67	66	57	65	57	55	58	65	51
59	70	63	59	74	71	58	51	43	44
61	59	61	70	61	63	54	60	60	71

2. Compute the third quartile, second quintile, and the ninth decile of the following population of 26 daily expenses:

44.98	66.56	49.28	68.03	54.95	58.49	54.68
51.37	39.06	50.05	47.57	61.20	72.60	77.11
41.24	47.97	71.35	42.91	65.03	57.54	43.54
69.17	45.39	38.89	59.97	36.43		

3. Compute the third quartile, second quintile, and the ninth decile of the following population of 45 male models' heights, in inches:

71	72	70	71	72	72	70	70	71	70
73	72	70	72	70	74	72	72	73	71
71	72	71	73	72	71	72	71	74	71
71	71	74	73	72	71	72	70	71	71
73	70	70	72	72					

4. Compute the third quartile, second quintile, and the ninth decile of the following population of 15 students' attention spans, in minutes:

10.9	10.1	9.2	13.6	9.9	9.7	13.8	13.0
5.0	6.6	8.6	14.0	8.5	9.9	10.1	

5. Compute the third quartile, second quintile, and the ninth decile of the following population of 30 automobile gasoline mileages:

21.9	38.6	34.7	31.2	17.2	18.6	28.4	30.3
35.7	21.9	23.1	38.8	38.0	37.3	16.2	22.7
28.3	22.4	24.0	15.6	31.5	22.7	24.4	16.9
26.6	21.8	30.6	19.6	35.8	23.0		

6. Compute the third quartile, second quintile, and the ninth decile of the following population of 35 television repair bills:

60.33	55.82	55.99	165.58	96.48
154.74	137.06	108.82	67.42	79.94
87.44	159.72	88.50	175.32	165.41
153.53	104.00	178.08	49.35	65.01
119.59	121.20	130.35	130.98	154.85
155.94	124.60	52.77	75.76	94.34
126.01	158.76	107.22	111.72	98.59

7. Compute the third quartile, second quintile, and the ninth decile of the following population of 15 rats' weight gains, in grams:

8.8	28.3	9.6	9.6	42.0	43.3	5.3	19.6
34.2	13.2	4.2	45.9	5.9	12.6	10.7	

8. Compute the third quartile, second quintile, and the ninth decile of the following population of 32 executive salaries:

128,000	145,000	78,000	136,000	60,000	72,000
91,000	217,000	46,000	191,000	85,000	127,000
139,000	134,000	182,000	104,000	173,000	177,000
197,000	213,000	149,000	112,000	133,000	145,000
101,000	155,000	167,000	143,000	213,000	88,000
149,000	173,000				

9. Compute the third quartile, second quintile, and the ninth decile of the following population of 60 pocketsful of change, in cents:

7	18	0	89	4	24	6	21	21	15
19	45	55	71	32	29	54	62	33	38
11	79	69	51	86	76	60	63	28	72
0	32	57	18	85	27	63	48	58	37
86	25	15	74	5	21	62	26	68	1
87	81	84	71	1	78	83	30	8	69

10. Compute the third quartile, second quintile, and the ninth decile of the following population of 30 flying times (in minutes) to cross the United States:

334	317	290	289	285	264	294	280	324	301
306	316	286	328	358	276	358	308	325	354
289	281	253	268	328	270	268	318	289	332

11. Compute the third quartile, second quintile, and the ninth decile of the following population of 33 measurements of galvanic skin response, in volts:

0.0213	0.0182	0.0201	0.0188	0.0204
0.0181	0.0190	0.0198	0.0214	0.0204
0.0208	0.0206	0.0210	0.0206	0.0200
0.0215	0.0206	0.0199	0.0206	0.0210
0.0190	0.0192	0.0202	0.0218	0.0181
0.0217	0.0180	0.0180	0.0182	0.0201
0.0197	0.0190	0.0214		

CHAPTER 1 QUIZ

1. Which of the following is *not* part of constructing a histogram?
   a. intervals
   b. classes
   c. tallies
   d. medians
   e. relative frequencies
   f. areas

2. When we construct the bars of a histogram, the relative frequency (percentage) of the class is proportional to the _____ of the bar.

3. (True-False): The mean, median, mode, and 75th percentile are *all* measures of location of a population.

4. Under what circumstance is the median a preferable measure of location to the mean?

ANSWERS TO CHAPTER 1 EXERCISES

1.

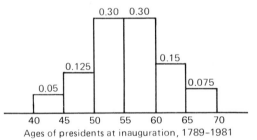

Ages of presidents at inauguration, 1789–1981

2. $\bar{x} = 54.825$; $\Sigma x_i = 2193$

3. Example 1: After ranking, the 10th smallest is 39, and the 11th smallest is 41. So the median = (39 + 41)/2 = 40 years.

   Example 2: The 20th smallest age is 55; 21st smallest age is 55; therefore, the median is 55.

4. Example 1: First Quartile $= Q_1 = $ 25th percentile
$$= \frac{27 + 30}{2} = 28.5$$

   Fourth Quintile $= QN_4 = $ 80th percentile $= 54$

   Seventh Decile $= D_7 = $ 70th percentile
$$= \frac{45 + 52}{2} = 48.5$$

   Example 2: $Q_1 = 51$; $QN_4 = 60.5$; $D_7 = 57$

5. There are four modes; the ages 51, 54, 55, and 57 each occur four times.

6. There are two modal classes: 50–54 and 55–59.

$$\text{50th percentile} = 55 + \frac{1}{12}(60 - 55) = 55.417$$

7. 0.05   less than 45
    0.175   less than 50
    0.475   less than 55
    0.775   less than 60
    0.925   less than 65
    1.000   less than 70

8.

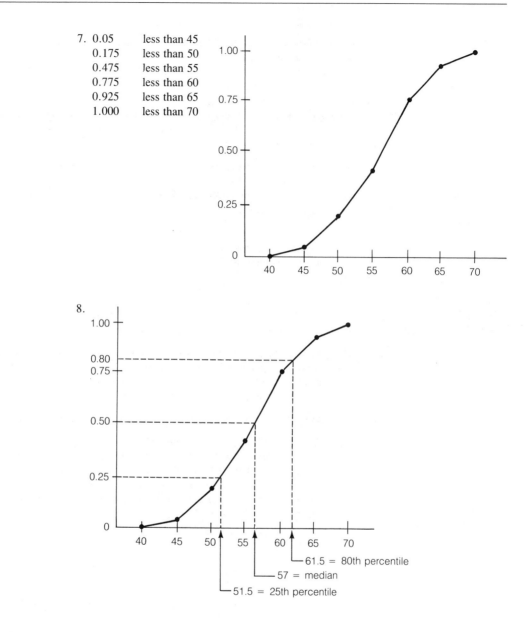

-61.5 = 80th percentile

-57 = median

-51.5 = 25th percentile

**ANSWERS TO
QUIZ (page 8)**

1. a. No. *Usually*, a population is a collection of numbers.
   b. No.
   c. No.
   d. Right.

2. a. Right.
   b. No. A population is a *collection* of numbers.
   c. No. The median is the halfway point in a population.

3. a. No. You forgot to divide the total sum by the number of numbers: 4.
        $60 \div 4 = 15$, the sample mean.
    b. No. Did you divide by 3? You're supposed to divide by the *number* of numbers; in this
        case there are four numbers: $60 \div 4 = 15$, the sample mean.
    c. Right: $60 \div 4 = 15$.
    d. The sample mean is the average value. You compute it by adding up all the numbers
        $(12 + 18 + 16 + 14) = 60$ and then dividing by the number of numbers (4):
        $60 \div 4 = 15$, the sample mean.

4. a. No. Sample statistics are numbers obtained from a sample of a population, not from
        the whole population.
    b. No. Histograms are pictures. The mean, maximum, minimum, and range are numbers
        computed from the whole population; they are population parameters.
    c. Right.

5. a. No. Histograms are pictures.
    b. No. Statistics are numbers.
    c. Right.

ANSWERS TO
CHAPTER 1 QUIZ

1. d. The median is the 50th percentile, or place where half the population is larger and half
        is smaller. It is not used in constructing a histogram.

2. Area. If you said "height," you are only correct when all the bars have the same width
    (that is, have uniform class intervals). If the bars are not of the same width, you must
    adjust the height of the bar accordingly, so that the area of the bar is proportional to the
    relative frequency of the class. (See page 15.)

3. True. If you think about it, the 75th percentile is *also* a measure of the location of the
    population.

4. The median is preferable to the mean as a measure of the location of a population when
    there are some extreme values in the population, which would distort the mean more than
    they would the median. For example, in a city of population 10,000, one billionaire would
    make the *mean* net worth of its citizens at least $100,000, even if everyone else was
    poverty-stricken.

Descriptive Statistics: Part 2

Terms we'll be learning about in this chapter

binomial distribution

distribution

normal distribution

normal table

sample

standard deviation

variance

z-score

2-1 VARIATION (VARIANCE)

Figure 2-1 shows the histograms of two populations. These two populations do not differ in *location*, but they differ in their *spread*. By spread we mean the amount the elements vary.

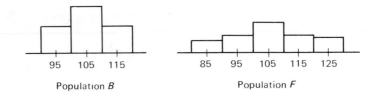

Population *B*   Population *F*

Figure 2-1

The elements of population *B* vary less than the elements of population *F*. But exactly what do we mean when we say they vary? To answer this question, we must talk about sampling.

SAMPLING

Imagine that we have written down each element (number) in a population on a slip of paper, one slip for each element. We put all the slips of paper in a hat, stir them around, and then reach in and draw one out. By this process, we are selecting an element (we say "drawing a sample of size 1") from the population. If we draw a second slip from the hat without putting the first slip back, we are *sampling without replacement*. If we return the original slip to the hat before drawing the second time, so that we have the possibility of drawing the same slip a second time, we are *sampling with replacement*. In this book we will usually be sampling with replacement.

When we sample elements from the population, we can ask ourselves if the elements change much from sample to sample. In the case of population *B* in Figure 2-1, we can see from the histogram that the most the elements vary is 30 units (from 90 to 120), and approximately 10 on average. With population *F*, the most the elements vary is 50 (from 80 to 130), and about 15 on average. The degree to which the elements in a population vary when sampled is of fundamental importance in statistics, both as a descriptive tool and as a predictive tool.

LOW VARIABILITY

Let's say you call a TV repairman and ask him for an estimate of the bill. If he tells you that his average repair bill is $50, his repair record could look like the one in Figure 2-2, or it could look like the one in Figure 2-3.

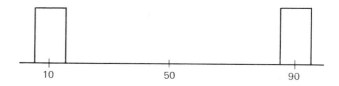

Figure 2-2   Repair record *A*.

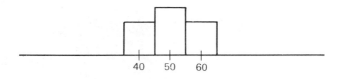

Figure 2-3   Repair record *B*.

Record *A* and record *B* both have mean 50, but they differ in their *variability*. Most people would prefer to use a repairman whose record was *B*, because the low variability means consistency—you know in advance pretty much what you're going to be paying. With record *A*, you may be in for a surprise: either a good surprise (a $10 repair bill) or a bad surprise (a $90 repair bill). So low variability corresponds to consistency and the ability to predict—or, as we say in statistics, *estimate*—with greater accuracy. The lower the variability of our "estimator," the better the estimate. We'll talk much more about this in Chapter 3.

"Joe's High Variability Repair Service is always happy to serve."

BUYING INSURANCE

Insurance exists because of people's desire for low variability. They wish consistency, an ability to plan for the future. For example, let's say car insurance will cost me $400 per year. According to the contract, if my car is demolished, it will be replaced by the insurance company at a cost to me of $100. That my car will be demolished is unlikely, occurring, say, with *probability* 0.02 in any one year. (This just means that there is a 2% chance that it will happen.) Let's also assume that whether or not I am insured, my other car expenses remain the same, at $1000 per year (for gas, oil, tires, and so on). If I don't take the insurance, then with probability 0.98 (98% of the time), my total costs for operating the car will be $1000 (no accident). But 2% of the time I *will* have an accident, forcing me to replace the car at a cost of $8000. In that unfortunate event, my cost for operating the car that year will be $8000 + $1000 = $9000. I can summarize that information in a special kind of histogram called a **distribution** (see Figure 2-4).

But what if I do buy insurance? I incur the extra cost of $400 each year, bringing my total costs of operating the car to $1400. The chances are still 98% that I will not demolish my car, and 2% that I will. But now in the event that I do, my total replacement cost is the $100 that the insurance company will charge me. The insurance company will pay the remainder of the cost of the car. This leads to the histogram shown in Figure 2-5.

**Figure 2-4 Distribution of costs of operating the car
without insurance.**

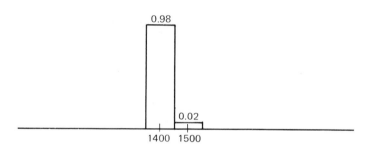

**Figure 2-5 Distribution of costs of operating the car
with insurance.**

What people are paying for when they buy insurance is the ability to "squeeze" all the outcomes close together, so that they can plan for their financial future. (As a side issue, let's note that if we compute the means of these two distributions, we will be computing our costs on the average, just as we computed means earlier.) Without insurance the mean is 0.98($1000) + 0.02($9000) = $1160. With insurance the mean is 0.98($1400) + 0.02($1500) = $1402. As you would expect, it costs you, on the average, to buy insurance. But most people do it anyway, because they are willing to pay for low variability. Statisticians would say that buying insurance was buying lowered variation and paying for it with decreased expectation. The means we computed of $1160 without insurance and $1402 with insurance are called **expected values**, but don't worry about them for now.

BUYING A DRESS SHOP

Let's say you were going to buy a dress shop. After inspecting the books, you concluded that the shop had, on the average, 10 customers per hour. Your staffing needs would be very different depending on which of the two histograms, Figure 2-6 or Figure 2-7 (on page 44), described the number of customers.

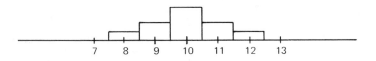

Figure 2-6   Number of customers in a one-hour period.

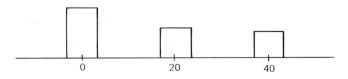

Figure 2-7   Number of customers in a one-hour period.

Again, the issue is variability—how much the elements in the population vary. If the distribution of customers is described by Figure 2-6, you can plan your staffing needs more easily than if the distribution is given by Figure 2-7. This points up the fundamental principle: *the lower the variability, the greater the predictive capability.*

THE VARIANCE OF A POPULATION

When we first think of measuring variability, we think of a population parameter we have already introduced: the *range*. As you may recall from Chapter 1,

$$\text{Range} = \text{Maximum} - \text{Minimum}.$$

But the range suffers from the defect that while it tells us the *maximum* variability in the elements of a population, it fails to account for *how frequently* this happens.
Consider the following three populations:

Population C:	0	0	0	0	0	10	10	10	10	10
Population D:	0	4	4	4	4	6	6	6	6	10
Population E:	0	5	5	5	5	5	5	5	5	10

Notice that all three populations have the same range ($10 - 0 = 10$), but population C tends to vary more than population D, and population D varies more than population E. To see that this is true, think of this: when we sample from E, we get 5 most of the time, whereas with population C, about half the time we get 0 and about half the time we get 10. So you can see that there's more varying (hence more variation) in population C than in D or E. I have drawn the means of the three populations

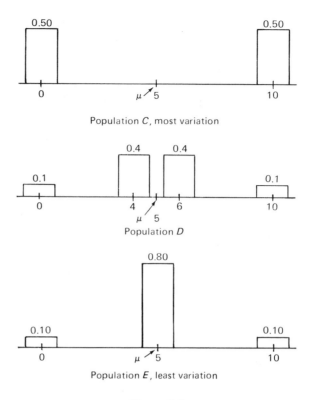

Figure 2-8

(Figure 2-8) and labeled the mean μ in each case (μ is the Greek letter mu and is frequently used to stand for the population mean).

Now let's try to define some notion of variability that has the variation of population C more than that of population D, and D more than E. (Looking at the histograms in Figure 2-8, you can see that the variability of C should be "large," while that of D and E should be "small." Study the three histograms until you can see it.) Although the range is 10 in each of the populations, the elements in populations D and E don't assume those extreme values *very often*. In population D, most of the elements are 4s and 6s, which have slight variability. In E, most of the elements are 5s, which have no variability.

Looking at the three histograms in Figure 2-8, can you see that one measure of variability that captures our intuitive picture is *how much* the elements vary from the mean μ, and *how often*? (Read this sentence again, look at the histograms, and take your time.) In population C, the elements are at distance 5 from the mean *all the time*. In population D, only 20% of the elements are at distance 5 from the mean. This means that on average, 20% of the time that elements are sampled, they will be found to be at distance 5 from the mean, while 80% of the time the elements are at distance 1 from the mean. In population E, 20% of the elements are again at distance 5 from the mean, but the remaining 80% are at the mean, or at distance 0 from the mean.

So, we're beginning to evolve a numerical measure of variability. It has to do with *how far* the elements of the population lie from the mean, taking into account *how often*. If the elements in a population of size 100 were labeled $x_1, x_2, x_3, \ldots, x_{100}$, then

$$\frac{\sum_{i=1}^{100} (x_i - \mu)}{100} \qquad (1)$$

would be a weighted average distance of the elements from the mean. Unfortunately, some of these distances would be positive (when x_i is larger than μ) and some would be negative (when x_i is smaller than μ), and the total sum in (1) would be 0. To overcome this, we *square* each of the distances in (1) to obtain a *mean squared distance*:

$$\frac{\sum_{i=1}^{100} (x_i - \mu)^2}{100} . \qquad (2)$$

And this measure of variability (2) is called the **variance** of the population, denoted σ^2, or σ^2_{pop}.

Of course, we wrote (2) assuming there were 100 elements in the population. In general, if there are n elements in the population, then the variance is

$$\sigma^2 = \sigma^2_{pop} = \frac{\sum_{i=1}^{n} (x_i - \mu)^2}{n} . \qquad (3)$$

CALCULATING THE VARIANCE OF POPULATION C

Usually we'll use a calculator to compute variance. But to emphasize the point that variance is mean *squared* distance, let's take another look at population C (Figure 2-8). By observing (via the histogram) that the distance from the mean of the population μ to every element of the population is the same, namely, 5, we get

$$|10 - 5| = 5 \qquad \text{and} \qquad |0 - 5| = 5.$$

So the distance squared equals $5^2 = 25$ for every element of the population. And the mean (average) distance squared must be 25. Thus the variance of population C, σ^2_C, is 25.

> $|A - B|$ stands for the distance between the numbers A and B and is always positive or 0.

CALCULATING THE VARIANCE OF POPULATION *D*

To calculate the variance of population D, we first calculate each of the distances from μ to each of the 10 elements in the population. Then we square each of the distances and calculate the mean of the 10 distances squared (see Table 2-1). As you can see from the table, $\sigma_D^2 = 5.8$.

Table 2-1   Calculating the Variance of Population *D*

Element: x_i	Distance: $x_i - \mu$	(Distance)²: $(x_i - \mu)^2$
0	$0 - 5 = -5$	25
4	$4 - 5 = -1$	1
4	$4 - 5 = -1$	1
4	$4 - 5 = -1$	1
4	$4 - 5 = -1$	1
6	$6 - 5 = \ \ 1$	1
6	$6 - 5 = \ \ 1$	1
6	$6 - 5 = \ \ 1$	1
6	$6 - 5 = \ \ 1$	1
10	$10 - 5 = \ \ 5$	25

$$\Sigma(x_i - \mu)^2 = 58$$

$$\frac{\Sigma(x_i - \mu)^2}{n} = \frac{58}{10} = 5.8$$

Number of elements in population

$$\text{var}(D) = \sigma_D^2 = 5.8$$

CALCULATING THE VARIANCE OF POPULATION *D*
IF YOUR CALCULATOR HAS AT LEAST ONE MEMORY

If your calculator has at least one memory register (it does if it has keys marked STO and RCL , for "store" in memory and "recall" from memory), you can calculate the variance by calculating the deviations from the mean, then squaring each, and then *adding* each new squared deviation to a "running sum" of the squares, using the SUM key. The SUM key will add whatever is in the display to whatever quantity is in the memory. After you've added all the squared deviations to memory, you then recall the sum and divide by the number of elements in the population. Remember to start out the calculation by clearing the memory: CLx , STO . This puts 0 in the display and then stores it in the memory.

For population D, the calculation would go like this: $\boxed{\text{CLx}}$, $\boxed{\text{STO}}$ (now we have 0 in the memory), $\boxed{0}$, $\boxed{-}$, $\boxed{5}$, $\boxed{=}$, $\boxed{x^2}$ (now we've computed our first squared deviation), $\boxed{\text{SUM}}$ (this adds the squared deviation to the memory), $\boxed{4}$, $\boxed{-}$, $\boxed{5}$, $\boxed{=}$, $\boxed{x^2}$, $\boxed{\text{SUM}}$ (the second squared deviation, added to memory), $\boxed{4}$, $\boxed{-}$, $\boxed{5}$, $\boxed{=}$, $\boxed{x^2}$, $\boxed{\text{SUM}}$, and so on, through the tenth deviation. We now have the sum of the 10 squared deviations in memory. We press $\boxed{\text{RCL}}$, $\boxed{\div}$, $\boxed{1}$, $\boxed{0}$, $\boxed{=}$ to recall this sum to the display and divide by 10, the size of the population, to obtain the variance: 5.8.

CALCULATING THE VARIANCE OF POPULATION D IF YOUR CALCULATOR IS FOUR-FUNCTION

You'll need pencil and paper now, to construct a table similar to Table 2-1. Your calculator has no memory, so you can't calculate the squares and remember them (add them up) at the same time. So you need to calculate the squares one by one, enter them on the table as we did in Table 2-1, and then go back and calculate the mean of these squared distances by adding them up and dividing by 10, their total number.

EXERCISE 1

Calculate the variance of population E. As you may recall, this variance should be less than the variance of population D, which was 5.8. So fill in Table 2-2 and then use your calculator to recalculate the variance of population E.

Table 2-2 Calculating the Variance of Population E

Element: x_i	Distance: $x_i - \mu$	(Distance)^2: $(x_i - \mu)^2$
0	$0 - 5 = ($ $)$	25
5	$5 - 5 = ($ $)$	0
5	$($ $=$ $)$	$($ $)$
5	$($ $=$ $)$	$($ $)$
5	$($ $=$ $)$	$($ $)$
5	$($ $=$ $)$	$($ $)$
5	$($ $=$ $)$	$($ $)$
5	$($ $=$ $)$	$($ $)$
5	$($ $=$ $)$	$($ $)$
10	$($ $=$ $)$	$($ $)$

$$\Sigma(x_i - \mu)^2 = (\quad)$$

$$\text{var}(E) = \sigma_E^2 = \frac{\Sigma(x_i - \mu)^2}{10} = (\quad)$$

Remember: μ = mean of population E.

Did you get 5? That's the variance of population E. Compare your Table 2-2 with mine, Table 2-3 on page 50.

TWO EXAMPLES OF POPULATION VARIANCE

Now that you know that variance is a measure of spread, think back to Examples 1 and 2 in Chapter 1, ages at coronation (kings and queens) and ages at inauguration (presidents). Which of these two populations would you expect to have greater variance?

If you look at the ages of the presidents at inauguration, you see that most of them are between 50 and 60 years old; intuitively you would expect this, because to become president requires years of political career-building. But whoever is first in line to the throne will become king or queen whenever the current ruler dies. Just when this happens is a bit more random than when one becomes president. So we would expect the variance of ages at coronation to be greater than the variance of ages at inauguration. In fact, we get

$$\text{var(population of Example 1)} = \sigma^2_{\text{kings}} = \frac{\sum (x_i - \mu)^2}{n}$$

$$= \frac{\sum (x_i - 40.4)^2}{20}$$

$$= 175.44$$

and

$$\text{var(population of Example 2)} = \sigma^2_{\text{pres}} = \frac{\sum (x_i - \mu)^2}{n}$$

$$= \frac{\sum (x_i - 54.825)^2}{40}$$

$$= 36.75.$$

THE VARIANCE OF A SAMPLE

As we mentioned in the introduction, there are times when we want information about (some parameter of) a population, but all we have is a sample from that population. For example, imagine we are scientists studying a new chemical, and we

Table 2-3   Calculating the Variance of Population *E*

Element: x_i	Distance: $x_i - \mu$	(Distance)^2: $(x_i - \mu)^2$
0	$0 - 5 = -5$	25
5	$5 - 5 = \ \ 0$	0
5	$5 - 5 = \ \ 0$	0
5	$5 - 5 = \ \ 0$	0
5	$5 - 5 = \ \ 0$	0
5	$5 - 5 = \ \ 0$	0
5	$5 - 5 = \ \ 0$	0
5	$5 - 5 = \ \ 0$	0
5	$5 - 5 = \ \ 0$	0
10	$10 - 5 = \ \ 5$	$\underline{25}$

$$\Sigma(x_i - \mu)^2 = 50$$

$$\text{var}(E) = \sigma_E^2 = \frac{\Sigma(x_i - \mu)^2}{10} = 5$$

wish to determine which of three processes will produce the greatest amount of the chemical—process C, process D, or process E. We could repeat an experiment using each of the three processes 10 times, measuring the yield (in grams, say) of the chemical each time.

Yield (in grams)

Process C: 0	0	0	0	0	10	10	10	10	10
Process D: 0	4	4	4	4	6	6	6	6	10
Process E: 0	5	5	5	5	5	5	5	5	10

Now each of three sets of numbers above is not a population, but a *sample from a population*. The actual population of process C is the yields that would be obtained from *every* repeated application of that process. So it's really a "potential population" in some sense. All we can do is look at a sample from this population and make whatever judgments (inferences) we can about the population from the sample we obtain. We can never look at the entire population (because we cannot conduct every repetition of the process). Notice that we now have *samples* of three processes C, D, and E, and three *populations* C, D, and E, each of size 10.

We can talk about the variability of a sample, just as we discussed the variability of a population. There is one difference, however. When we compute the variance of a population, we measure deviations from the population mean. When we compute the variance of a sample, we also want to measure the deviations from

A typical sample:

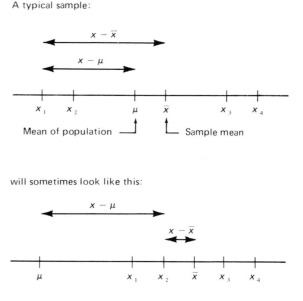

will sometimes look like this:

Figure 2-9

the population mean, but we usually don't know the mean of the population from which we are sampling. So we have to *estimate* the population mean μ (a parameter of the population) by the mean of the sample $\bar{x}$ (a statistic) and then use this mean to compute deviations and thus variance. The trouble with this approach is that it underestimates the variance. You can see this from Figure 2-9. Whenever the sample clusters about some point not μ, which will happen some percentage of the time, $|x_i - \bar{x}|$ will be much smaller than the "true" deviations $|x_i - \mu|$. So when we use $\bar{x}$ in place of μ to measure the deviations, we will underestimate population variance on the average, which would make sample variance a *biased estimator* (biased to the small side). To correct for this bias, we calculate the sample variance by dividing by $n - 1$ instead of by n in formula (3):

$$s^2 = \frac{\sum (x_i - \bar{x})^2}{n - 1}. \tag{4}$$

Note that we denote the sample variance by s^2 (instead of σ^2).

Recall our three chemical processes and the sample of 10 yields from each of them. Now let's compute the sample variance for process C, using formula (4) (see Table 2-4). Notice that the only difference in computing the variance of a *sample* and the variance of a population is that in one case we use the sample mean and divide by $n - 1$, formula (4), and in the other case we use the population mean and divide by n, formula (3). *It is very important to fix the two situations clearly in your*

Table 2-4 Calculating the Sample Variance for Process C

Element: x_i	Distance: $x_i - \bar{x}$	(Distance)^2: $(x_i - \bar{x})^2$
0	$0 - 5 = -5$	25
0	$0 - 5 = -5$	25
0	$0 - 5 = -5$	25
0	$0 - 5 = -5$	25
0	$0 - 5 = -5$	25
10	$10 - 5 = 5$	25
10	$10 - 5 = 5$	25
10	$10 - 5 = 5$	25
10	$10 - 5 = 5$	25
10	$10 - 5 = 5$	25

$$\frac{\Sigma(x_i - \bar{x})^2}{n-1} = \frac{250}{9}$$

$$= 27.78 = s^2$$

mind. Review formulas (3) and (4), for population variance and sample variance, respectively:

$$\sigma^2 = \frac{\sum_{i=1}^{n}(x_i - \mu)^2}{n} \tag{3}$$

$$s^2 = \frac{\sum_{i=1}^{n}(x_i - \bar{x})^2}{n-1} \tag{4}$$

EXERCISE 2

Use your calculator to compute the sample variance for process D.

_____

EXERCISE 3

Use your calculator to compute the sample variance for process E.

_____

THE STANDARD DEVIATION

Notice that while the variance measures the spread of the elements in either a sample or a population, the variance tends to be much larger than the average spread of the elements. That's because when we compute the deviations, we *square* them before averaging them. So while the variance is a good measure of spread, its scale is not a useful one (because it tends to be much larger than the spread most of the time). But we can easily overcome this problem by taking the *square root of variance* and using it as our measure of spread. And this is exactly what we shall do, for both population variance and sample variance. The square root of variance is called **standard deviation**, denoted by σ (the Greek letter sigma) in the case of population standard deviation, and denoted by s in the case of sample standard deviation:

$$\sigma = \text{Population standard deviation} = \sqrt{\frac{\sum (x_i - \mu)^2}{n}} \tag{5}$$

$$= \text{Square root of formula (3),}$$

$$s = \text{Sample standard deviation} = \sqrt{\frac{\sum (x_i - \bar{x})^2}{n - 1}} \tag{6}$$

$$= \text{Square root of formula (4).}$$

If your calculator doesn't have a square root key, use this trial-and-error method: To compute $\sqrt{6000}$ to two decimal places, make an initial guess. Since $8^2 = 64$, then $80^2 = 6400$. So $\sqrt{6000}$ should be a little less than 80—say, 75. Calculate 75^2 (use $\boxed{\times}$, $\boxed{=}$ if necessary) $= 5625$. It's too small and 80 was too big, so try 77: $77^2 = 5929$. Since 77^2 is close to 6000, 78^2 will probably be too much. Guess 77.5: $77.5^2 = 6006.25$, just a little too big. Try 77.4: $77.4^2 = 5990.76$, too small. So try 77.45: $77.45^2 = 5998.5$. This *may* be the correct answer, but try 77.46: $77.46^2 = 6000.05$. This *must* be the correct answer, to two places.

You can judge how standard deviation measures spread by plotting sample standard deviation ($1s = 1\sigma_{n-1}$) against the histogram of process C, and population standard deviation ($1\sigma = 1\sigma_n$) against the histograms of populations B and F (see Figure 2-10).

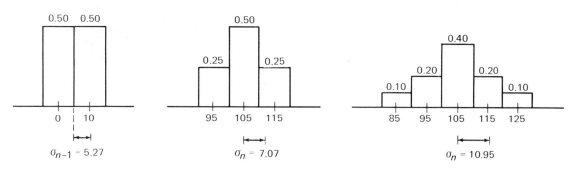

$\sigma_{n-1} = 5.27$

Process C

$\sigma_n = 7.07$

Population B

$\sigma_n = 10.95$

Population F

Figure 2-10

EXERCISE 4   Calculate the *sample* standard deviations of processes C, D, and E.

$$s_C = \text{\underline{\hspace{5cm}}}$$

$$s_D = \text{\underline{\hspace{5cm}}}$$

$$s_E = \text{\underline{\hspace{5cm}}}$$

USING A CALCULATOR WITH BUILT-IN STATISTICAL FUNCTIONS

So far I've shown you how to compute standard deviations of populations and samples with a calculator with or without memory. Many advanced-generation calculators have built-in functions that calculate standard deviations automatically. On the TI-35 you will observe two keys, $\boxed{\sigma_n}$ and $\boxed{\sigma_{n-1}}$. Here is how these keys work: first you key in the data, whether a population or a sample, by means of the $\boxed{\Sigma+}$ key. Let's use the data of process C as an example:

Process C: 0   0   0   0   0   10   10   10   10   10

Clear the statistical registers. Then key in each of the 10 numbers, using the $\boxed{\Sigma+}$ key. The calculator should be displaying ___ ***10*** because you keyed in 10 numbers. Now press $\boxed{2nd}$, $\boxed{\sigma_{n-1}}$ (because this is a sample from a population,

not a complete population). The calculator should be displaying 5.2704628 , which is the *sample standard deviation*. Recalling that variance is the square of standard deviation, now press $\boxed{x^2}$ to compute the sample variance. Your calculator should now be displaying 27.777778 (see Figure 2-11). [*Warning*: On the TI-35 and some similar calculators, in order to perform two-number functions (such as + and ÷) after you have used the $\boxed{\Sigma+}$ key or some other statistical feature of the machine, it is necessary to clear statistical registers ($\boxed{\text{CSR}}$). Of course, $\boxed{x^2}$, or squaring, is a one-number function, so clearing is not necessary.]

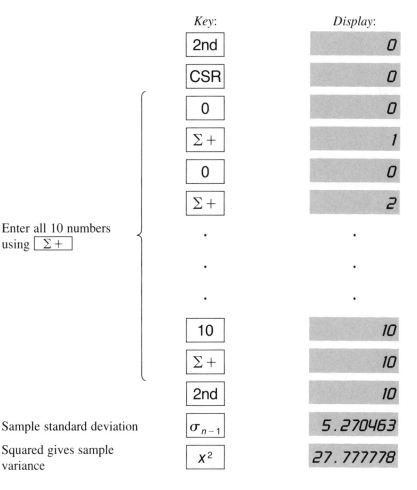

	Key:	*Display*:
	2nd	0
	CSR	0
	0	0
	$\Sigma+$	1
	0	0
	$\Sigma+$	2
Enter all 10 numbers using $\boxed{\Sigma+}$	.	.
	.	.
	.	.
	10	10
	$\Sigma+$	10
	2nd	10
Sample standard deviation	σ_{n-1}	5.270463
Squared gives sample variance	x^2	27.777778

Figure 2-11

EXERCISE 5a

Use your $\boxed{\Sigma+}$ key and $\boxed{\sigma_{n-1}}$ to calculate the sample standard deviations of processes D and E. After you obtain the sample standard deviation, calculate the sample variance by squaring.

Process D: $\sigma_{n-1} = $ _____

$s_D^2 = $ _____

Process E: $\sigma_{n-1} = $ _____

$s_E^2 = $ _____

To calculate the population standard deviation and variance, we use the $\boxed{\sigma_n}$ key instead of the $\boxed{\sigma_{n-1}}$ key. [Or if your calculator has no statistical functions, use formula (5) instead of (6).] It is not necessary to reenter the data; simply press $\boxed{\text{2nd}}$, followed by $\boxed{\sigma_n}$ rather than $\boxed{\sigma_{n-1}}$. The calculator will then be displaying the population standard deviation. To obtain the population variance, we square the result, pressing $\boxed{x^2}$ to obtain it.

EXERCISE 6a

Calculate the variances for populations D and E by the same method discussed above.

$\sigma_D^2 = $ _____

$\sigma_E^2 = $ _____

THE SHORTCUT FORMULA FOR VARIANCE AND STANDARD DEVIATION

If your calculator doesn't have statistical registers or a $\boxed{\Sigma+}$ key, then you have to compute variance (and standard deviation) by making "two passes" at the data. First you must go through the data to calculate the mean, and then you must go through the data a second time to calculate deviations from the mean. If the population or sample whose variance you are calculating is small, this really isn't a problem. But if the sample size is in the hundreds (or larger), making a second pass can be quite a chore. Fortunately, it's not necessary to make a second pass at the data if you can calculate *two* running sums: Σx_i and $\Sigma(x_i^2)$. The first of the two quantities you are calculating is the sum of all the elements in the sample or population; the second quantity is the sum of the *squares* of those elements (which is *not* the same as the square of the sum of the elements). You may need a few minutes to see if your calculator can handle this. You'll need a calculator with at least one memory register. If your calculator has no memory register, this shortcut method isn't for you.

Now I'll show you how to calculate $1 + 2 + 3 + 4$ and $1^2 + 2^2 + 3^2 + 4^2$ using a calculator with one memory register we can store in ($\boxed{\text{STO}}$) and add to ($\boxed{\text{SUM}}$), and which we will later recall ($\boxed{\text{RCL}}$). Work along with me on your calculator, and refer to Figure 2-12 to check it. We clear our arithmetic memory by

Key	Display	In Memory
CLR	0	?
STO	0	0
+	0	0
1	1	0
SUM	1	1
x^2	1	1
=	1	1
+	1	1
2	2	1
SUM	2	3
x^2	4	3
=	5	3
+	5	3
3	3	3
SUM	3	6
x^2	9	6
=	14	6
+	14	6
4	4	6
SUM	4	10
x^2	16	10
=	30	10
RCL	10	10

Figure 2-12

pressing CLR , STO , which puts 0 in the accumulator and stores 0 in the memory (see Figure 2-12). Then we key in 1 , SUM , x^2 . We now have 1 stored in memory and 1^2 in the accumulator. Now we press + , 2 , SUM , x^2 , = . When we pressed SUM , the ⬛ 2 in the display was added to the 1 in memory, giving 1 + 2 in memory and leaving ⬛ 2 in display. When we pressed x^2 , the ⬛ 2 in the display was squared, giving ⬛ 4 , and then = activated the earlier + we had keyed in, now displaying $1^2 + 2^2 =$ ⬛ 5 . Now we press + , 3 , SUM , x^2 , = . We have stored 1 + 2 + 3 in memory and are displaying $1^2 + 2^2 + 3^2 =$ ⬛ 14 . Finally, we press + , 4 , SUM , x^2 , = . The accumulator should be displaying $1^2 + 2^2 + 3^2 + 4^2 =$ ⬛ 30 . When we press RCL , the calculator will display ⬛ 10 , which is 1 + 2 + 3 + 4. And that's how we compute Σx_i and $\Sigma(x_i^2)$.

Now that we know how to compute Σx_i and $\Sigma(x_i^2)$ in one pass at the data, we can compute either the population variance or the sample variance from this. The shortcut formula for population variance is

$$\sigma_{\text{pop}}^2 = \frac{n\Sigma(x_i^2) - (\Sigma x_i)^2}{n^2}, \tag{7}$$

where n is the population size. The shortcut formula for sample variance is

$$s^2 = \frac{n\Sigma(x_i^2) - (\Sigma x_i)^2}{n(n-1)}, \tag{8}$$

where n is the sample size. Of course, (7) and (8) can be used as shortcut formulas for standard deviation, by taking square roots.

EXERCISE 5b

Calculate the sample standard deviation of processes D and E by means of formula (8). Remember to take square roots to get the final answer.

Process D: $\sigma_{n-1} =$ _____

   $s_D^2 =$ _____

Process E: $\sigma_{n-1} =$ _____

   $s_E^2 =$ _____

EXERCISE 6b

Calculate the variance of populations D and E by means of formula (7).

$\sigma_D^2 =$ _____

$\sigma_E^2 =$ _____

2-2 SAMPLING DISTRIBUTIONS

So far we have used histograms only to describe populations. But we shall also be using them to describe statistics. As you may recall, *a statistic is a number based on a sample drawn from a population*. For example, we could draw a sample of size 10 from the population of annual rainfall in Philadelphia and then compute the *maximum* of the 10 numbers in the sample. This would be the *sample maximum* and would be a statistic of the sample. Some very useful sample statistics that we already know are the *sample mean* and *sample variance*.

Just as we used histograms to describe the elements of a population, we can use histograms to describe the *possible* values of any sample statistic. We can think of the histogram of the statistic as a recording (on a percentage basis) of the results of drawing samples (of the one fixed size, say 10) from the population thousands of times and computing the value of the sample statistic each time. The histogram that records each value observed, together with its relative frequency, is the histogram of the statistic and is called the **distribution of the statistic**. So **distribution** is just another name for histogram for the special case of a sample statistic.

For example, suppose we have a certain population *P*. In the histogram for *P* (shown in Figure 2-13), 25% of the elements are between -0.25 and 0.25, 50% of the elements are between 0.75 and 1.25, and 25% of the elements are between 1.75 and 2.25. But suppose, for simplicity, 25% of the population *P* are 0s, 50% of the population 1s, and 25% of the population are 2s. Now suppose we draw out a sample of size 2 from population *P* and compute the mean of this two-element sample. This number, the mean of a sample of size 2, is a statistic. As such, we can describe the possible values that the statistic can take, along with *how often* these values are achieved.

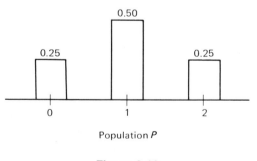

Population *P*

Figure 2-13

Notice that both Figures 2-13 and 2-14 are histograms. When we know the histogram for a population, we can mathematically calculate from it the histogram of any sample statistic from that population. Don't concern yourself here with *how*

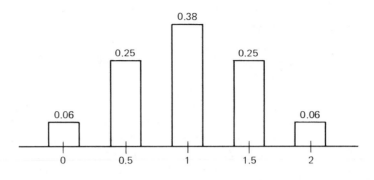

Figure 2-14   Distribution of the statistic $\bar{x}_2$, where $\bar{x}_2$ is a symbol for means of samples of size 2.

we do that; just assume that the calculation has been made for us, and we are now looking at the histogram (distribution) of the statistic $\bar{x}_2$ (Figure 2-14). What this distribution tells us is that when you draw a sample of size 2 from population P and compute the mean of that sample, 6% of the time that mean will be 0, 25% of the time it will be 0.5, 38% of the time it will be 1, and so on. Notice that the distribution of the statistic doesn't (*can't*) tell us what the value of the statistic is going to be this time, or the next time we draw a sample. But it can (and does) tell us on a percentage basis how likely each possible outcome is. Sometimes the outcomes are so numerous—instead of the possible outcomes being 0, 0.5, 1, 1.5, 2, the outcomes might be *all* numbers between 0 and 2, and there are infinitely many of those—that we don't say how likely each outcome is, but say instead how likely it is that the statistic will take a value *between* 0 and 0.5, or between 0.7 and 0.8. These histograms are called **continuous histograms**, and we'll be talking about them next.

2-3 CONTINUOUS HISTOGRAMS

When we use a histogram to describe a population or a sampling distribution, we can make our class interval widths as large as or as small as we like. For the cash register receipts (Chapter 1, page 22), it makes little sense to choose a class interval width representing less than 1¢, because the entire population of sales is given in increments of cents, and so the population is located at the integer (whole-number) values. In fact, if we choose interval widths of 1¢, our histogram would be mostly blank, because there are only 10 register receipts. See Figure 2-15, which shows *a part* of the histogram for the 10 register receipts. In cases like this, it makes no sense to continue to subdivide intervals to be smaller than 1¢ (and, indeed, you can see that even 1¢ is too small to adequately represent what's going on). However, when a population is *not discrete* (that is, where it can take on *any* value between

Figure 2-15

two whole numbers), such as in the rainfall example (which was reported to tenths of an inch, but where the rainfall could possibly be 35.5372653281 inch), you can subdivide to smaller and smaller class interval widths and the histograms continue to make sense. In many cases, as you subdivide more and more, the histograms will approach a *limiting histogram*, which will look like a smooth curve. This curve, or *continuous histogram*, while not technically a histogram, still gives what we call a "distribution," either a population distribution or a sampling distribution. Figure 2-16 illustrates this concept, with four histograms for the *same* population. Recall that the *area* of the bars is proportional to relative frequency.

As I said, we arrange the bars so that the area of each bar represents the proportion of time that the values in that class (interval) occurred. So the areas inside the bars in Figure 2-16a, b, c are all 1 square unit of area, because the total proportion of time that something from the histogram is selected is 100% of the time. Now we move on to Figure 2-16d. It's pretty clear that the histograms in Figure 2-16a, b, c tend to the curve of the distribution in Figure 2-16d. But what does such a continuous distribution mean?

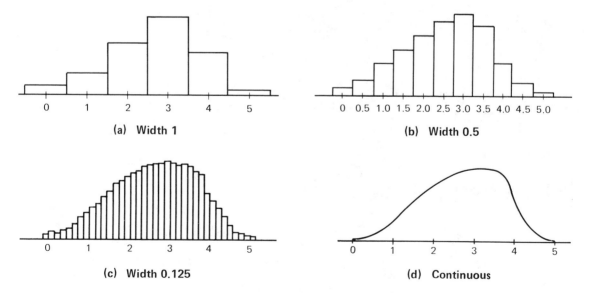

(a) Width 1

(b) Width 0.5

(c) Width 0.125

(d) Continuous

Figure 2-16

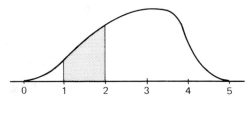

Figure 2-17

The shaded area between 1 and 2 on the continuous histogram in Figure 2-17 can be thought of as many tiny rectangles (see Figure 2-16c). As such, the area of each of those skinny rectangles is the proportion of time that the value of the statistic or the element drawn from the population falls in the corresponding class interval. The sum of all those proportions is the proportion of time that the value of the statistic or the element drawn from the population lies between 1 and 2. Thus, we have an interpretation of the shaded region. *Its area is exactly the proportion of time that the value of the statistic or the element drawn from the population will lie between 1 and 2.*

A virtue of the continuous distribution is that you may make the same interpretation between any two values, not just between integers. Thus, in Figure 2-18, the shaded region represents the proportion of elements in the population that lies between 1.3 and 2.7, just as in Figure 2-17 the shaded region represents the proportion of the population lying between 1 and 2. The vertical scale is always chosen so as to make the total area of the histogram 1 square unit.

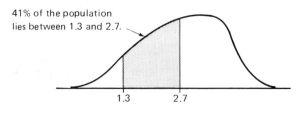

41% of the population
lies between 1.3 and 2.7.

Figure 2-18

2-4 THE NORMAL DISTRIBUTION

Two hundred years ago, scientists found that the discrepancies they observed while trying to record precise measurements always seemed describable by means of a distribution that had the same shape, regardless of what they were measuring. This was quite remarkable, since the measurements could be of temperature in one case and weight in another—and still the distribution of discrepancies (from the mean value) would appear to have this same "bell shape."

Now look at Figures 2-19, 2-20, and 2-21. Figure 2-19 shows the distribution

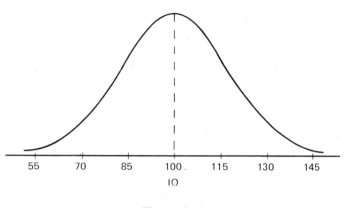

Figure 2-19

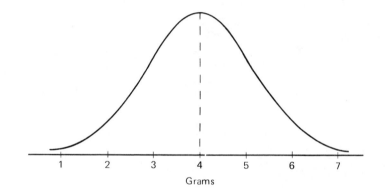

Figure 2-20

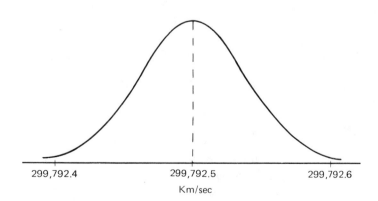

Figure 2-21

of IQs (from intelligence tests) of all school-age children in the United States. Figure 2-20 shows the distribution of the weight (in grams) of oak leaves from all oak trees. And Figure 2-21 shows the distribution of 152,488 measurements of the speed of light (in kilometers per second). The fact that all three figures have the same distribution is again remarkable, because one curve has to do with a population of IQ scores, another with the population of oak leaves, and the third is not derived from a population of objects at all, but rather from the repeated measurement of a single physical constant.

Sometimes these bell-shaped curves do not appear the same because the x-axis (horizontal axis) is either "stretched out" or "shrunk." Figures 2-22 and 2-23 show the same distribution of IQ scores represented in Figure 2-19; but the x-axis has been stretched out in Figure 2-22 and shrunk in Figure 2-23.

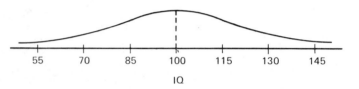

| 55 | 70 | 85 | 100 | 115 | 130 | 145 |

IQ

Figure 2-22

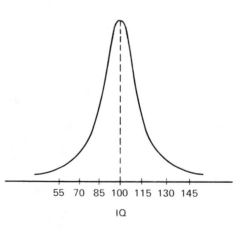

55 70 85 100 115 130 145

IQ

Figure 2-23

It was eventually proved by mathematicians that the bell-shaped curve, or **normal distribution**, as it is properly called, is also the approximate distribution for sample means, when the sample size gets larger and larger, for any population with finite variance. (This is the main reason we will be studying the normal distribution.) In practice it was found that for most populations, the normal distribution was already a good approximation for describing how sample means distribute themselves, if the sample size is 150 or more. Once we see that *all* the normal distributions are the same, except for where they are located (mean) and how stretched out they are (variance), we really need study only *one* normal distribution, the so-called standard normal distribution.

The **standard normal distribution** is the one that has mean 0 and standard deviation 1. To study the normal distribution that has mean 10 and standard deviation 5, we merely superimpose the same bell-shaped curve over a shrunken and relocated *x*-axis, as shown in Figure 2-24. Thus, to determine the area in the shaded part of the normal distribution with mean 10 and standard deviation 5, we simply look at the corresponding portion of the standard normal distribution and see what area it contains. (These areas are provided in a **normal table** of values for the stan-

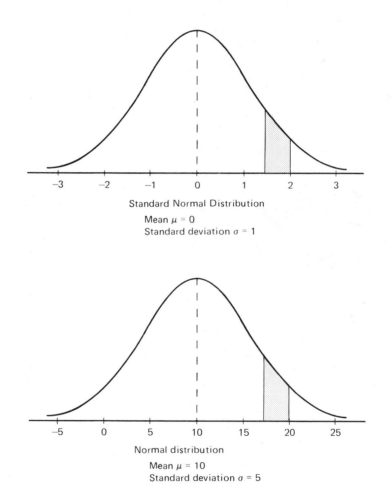

Figure 2-24

dard normal distribution, Table 1, at the end of the book.) To compute the area between 17.5 and 20 in the normal distribution with mean $\mu = 10$ and standard deviation $\sigma = 5$, we convert the numbers 17.5 and 20 into *z*-**scores**, or deviations
measured from the mean μ, measured in units of σ. Thus, 20 is 10 greater than the
mean of 10, and 10 is 2σ; so 20 is 2σ *greater than* the mean, giving 20 a *z*-score of
+2. Since 17.5 is 7.5 greater than the mean, and 7.5 measured in σ is 1.5σ (since
$\sigma = 5$), the *z*-score of 17.5 is +1.5 (see Figure 2-25). Thus, the area between 17.5
and 20 in this normal distribution is the same as the area between 1.5 and 2 in the
standard normal distribution. We can look up that value in Table 1.

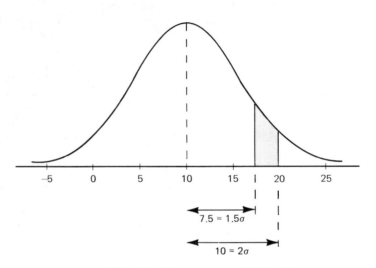

17.5 − 10 = 1.5σ, so *z*-score of 17.5 is 1.5.
20 − 10 = 2σ, so *z*-score of 20 is 2.

Figure 2-25

HOW TO COMPUTE A *z*-SCORE

Let X be a number in a given normal distribution. Then if μ is the mean of the
distribution and σ is the standard deviation of the distribution,

$$z = \frac{X - \mu}{\sigma}$$

is the *z*-score of X. In our example above, the *z*-score of 20 is

$$z = \frac{20 - 10}{5} = 2$$

and the *z*-score of 17.5 is

$$z = \frac{17.5 - 10}{5} = \frac{7.5}{5} = 1.5$$

(see Figure 2-26). Note that *z*-scores can be negative. Whenever *X* is smaller than μ, *X* falls to the *left* of μ, which corresponds to a *z*-score less than 0.

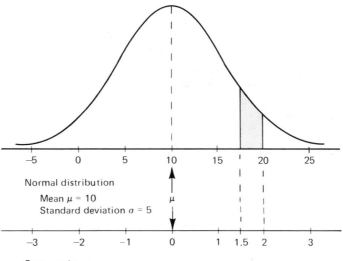

Figure 2-26

EXERCISE 7   A normal distribution is known to have mean 50 and standard deviation 12. Convert the numbers 50, 56, 68, 80, 44, and 38 to *z*-scores.

EXERCISE 8   IQ scores are observed to come from a normal distribution with mean 100 and standard deviation 16. Convert the following IQs to *z*-scores: 116, 132, 148, 100, 108, 104, 84.

EXERCISE 9   A normal distribution is known to have mean 100 and *variance* 400. What interval *about the mean* is $\pm 2\sigma$?

COMPUTING AREAS UNDER
THE NORMAL DISTRIBUTION

The shaded regions in Figure 2-24 both have the same area if the area under the curve in both cases is 1 square unit, or 100%. To determine what percentage of the total area the shaded region in Figure 2-24 is, we need a table of areas of the normal distribution. We know that a table of areas of the standard normal distribution will work for *all* normal distributions. If we had one table where we could look up the answer, the table would have to be indexed by both the left-hand end of the interval, which in this case is 1.5, and the right-hand end of the interval, which is 2. But any table with entries for both left- and right-hand endpoints would be very unwieldy. So instead, we arrange a table of values of just the right-hand endpoints. We always use the middle of the distribution, 0, as the left-hand endpoint. In Figure 2-27 the area under the standard normal distribution between 0 and 1.5 is 0.4332. In Figure 2-28 the area under the normal distribution between 0 and 2 is 0.4772. So, the area

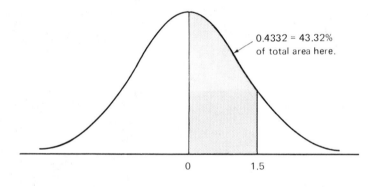

Figure 2-27

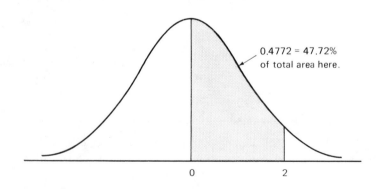

Figure 2-28

under the normal distribution between 1.5 and 2 is the *difference* between the areas of Figures 2-27 and 2-28. If we remove the area between 0 and 1.5 from the area between 0 and 2, we leave the area between 1.5 and 2 (see Figure 2-29). Thus, the area between 1.5 and 2 is $0.4772 - 0.4332 = 0.0440$.

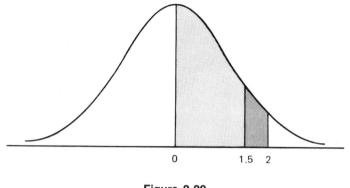

Figure 2-29

If we wish to compute an area that lies entirely to the left of 0, which involves negative z-scores, we use the fact that the normal distribution is *symmetric* about 0, reflecting the picture through the y-axis (the vertical line above 0) and making our computations with positive z-values (see Figure 2-30).

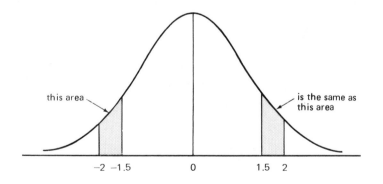

Figure 2-30   The normal distribution is symmetric.

If we wish to compute an area that straddles 0, such as the one shown in Figure 2-31, then we split the picture into the left half and the right half, using the symme-

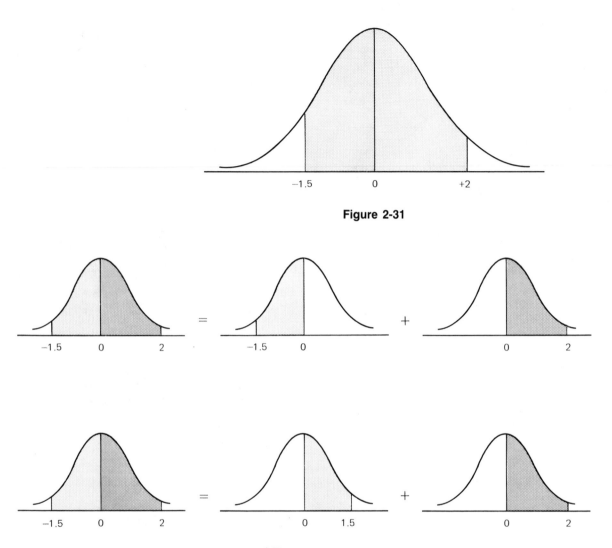

Figure 2-31

Figure 2-32

try of the normal distribution to determine the left half as if it contained positive values (see Figure 2-32). So the area under the normal distribution between -1.5 and 2 is equal to the area between 0 and $+1.5$ plus the area between 0 and $+2$, or $0.4332 + 0.4772 = 0.9104$.

What we need, finally, is a table of values of the area under the normal distribution between 0 and *every positive value*. This is what we are given in Table 1, "Areas of the Normal Distribution." A small portion of this table appears in Table 2-5. I have filled in only a portion of the table, to make it easier to talk about. Notice that the z-scores are arranged vertically in the first column of the table. Adjacent to

Table 2-5   From "Areas of the Normal Distribution"

z-score	0	1	2	3	4	5	6	7	8	9
1.5	0.4332	0.4345					0.4406			
1.6										
1.7										
1.8										
1.9										
2.0	0.4772			0.4788						0.4817

the z-score 1.5 is the value 0.4332. This means that the area under the normal distribution between 0 and 1.5 is 0.4332 (and, as I said, the area between 0 and −1.5, as well). Notice that this entry, 0.4332, is placed under the numeral 0. This 0 is actually the second decimal place of the z-score 1.5, so the z-score is actually 1.50. The number immediately to the right of 0.4332, 0.4345, is placed under the numeral 1, also in the row of 1.5. It is therefore the entry for 1.5<u>1</u>. We have omitted the remaining entries in the first row of our table, except for the entry under the numeral 6. Can you guess what this entry represents? (*Answer*: It is the area under the normal distribution between 0 and 1.5<u>6</u>.) All the entries in the 1.5 row would be the entries for z-scores of 1.50, 1.51, 1.52, and so forth, up to 1.59. The next entry in the table, the one for 1.60, would be found where? (*Answer*: In the second column of the table, directly below the entry for 1.50.)

Now that you know how to use the normal table, you can easily find the area under the normal distribution between 0 and 2. Look in the partial table (Table 2-5) to find it. (Did you find the entry 0.4772?) It's in the row marked 2.0 and the column marked 0, because it's the entry for 2.0<u>0</u>. The complete normal table is Table 1, found on page 294.

EXERCISE 10

What is the area between z-scores

a. 1.56 and 2.03?   b. −2.03 and −1.56?   c. −1.56 and 2.03?

a. _____

b. _____

c. _____

EXERCISE 11

A normal distribution is known to have mean 50 and standard deviation 10. What is the area between 65 and 70.3? (*Hint*: First convert to z-scores.)

_____

2-5 THE BINOMIAL DISTRIBUTION

Since we've spent so much time talking about the normal distribution, you probably think it's the only one we need to study. Unfortunately, that's not the case. A number of situations are not adequately described by a normal distribution: the proportion of heart attack victims who will survive for at least one year if given a certain new drug; the number of drivers stopped for a routine traffic inspection who fail to possess a valid driver's license; the number of questions a student will answer correctly on a 20-question multiple choice test, if he guesses at each answer. All three of these examples are instances of the *binomial distribution*. Because the binomial distribution occurs in such a wide variety of settings, and especially because it occurs in laboratory experiments (like the one testing the heart attack drug), it is an important distribution to study.

The **binomial distribution** is characterized by the following three properties:

1. There are a number of "trials" to the experiment, usually fixed in advance (the number of motorists stopped for inspection; the number of heart attack victims included in the study; the 20 test questions).

2. With each trial there is associated a likelihood, or probability, of "success" or "failure," which remains the same from trial to trial. Success can mean surviving the heart attack for one year or more, or possessing a valid driver's license (or *not* possessing a valid driver's license), or answering a test question correctly.

3. The trials are independent of one another; that is, what happens when the second motorist is stopped does not depend on what happened when the first motorist was stopped.

When we have an experimental situation satisfying these three properties, we call it an *independent trials process*, or *binomial process*. The only numbers we need to know to determine which binomial process we have are n, the number of trials, and p, the probability of success. By the way, just which outcome we call "success" and which we call "failure" is arbitrary. (When we know one, we can determine the other.) We could call the patients who do not survive one year with the drug "success" if we wished, and measure this probability instead. (However, this would probably be considered perverse.) From the Highway Patrol's point of view, success would probably be considered possession of a valid driver's license, even if what they really wished to measure was the proportion or percentage of drivers without valid licenses.

EXAMPLES OF THE BINOMIAL DISTRIBUTION

We've already talked about how histograms can describe the *possible* outcomes of some sampling process. Suppose we have a binomial process with $p = 0.3$ and $n = 6$. This means that we are going to run 6 trials of the binomial process (6 motorists stopped, or 6 heart attack victims treated, or 6 test questions asked), with probability of success 0.3 on each independent trial. The result of running this 6-trial experiment is that there will be a certain number of successes: either 0 successes, or 1 success, or 2, 3, 4, 5, or 6 successes. There are 7 possible outcomes in all. You can see, intuitively, that if the likelihood of success is only 0.3, then it's more likely on any trial of this experiment to get a failure than a success (the likelihood of failure is $1 - 0.3 = 0.7$), and so the likelihood of 6 successes (0 failures) is less than the likelihood of 0 successes (6 failures). So the 7 possible outcomes, from 0 successes to 6 successes, are not equally likely. In Figure 2-33 we have a "histogram" for the binomial process with $n = 6$, $p = 0.3$, describing the 7 possible outcomes of 6 binomial trials, each with probability 0.3 of success. (Actually, Figure 2-33 isn't exactly a histogram; there's no possibility of the value 1.5 or 1.7 occurring since all the possible outcomes are integers, or whole numbers.)

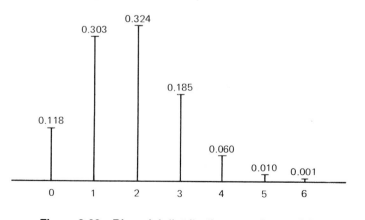

Figure 2-33   Binomial distribution, $n = 6$, $p = 0.3$.

The likelihood of each of the 7 possible values is indicated in the figure. These likelihoods represent what would happen if we repeated the experiment of 6 binomial trials (each with probability of success of 0.3) thousands and thousands of times. If we did that, of course, eventually some of the time we would observe every possibility, even 6 successes. The proportion of the time that we would observe 0 successes is given as 0.118, meaning that 11.8% of the time that we ran this 6-trial binomial experiment we would observe 0 successes. And 0.303, or 30.3%, of the time we would observe 1 success. And so forth. Notice that the sum of these proportions equals 1, or 100%. (Occasionally, the proportions will not *exactly* equal 1 because of round-off error. Don't worry about this if the numbers total close to 1.00.)

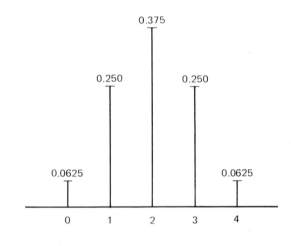

Figure 2-34   Binomial distribution, $n = 4$, $p = 0.5$.

Now take a look at the "histogram" shown in Figure 2-34. Note that this "histogram" of the binomial distribution is symmetric. This is because the probability of success and the probability of failure is the same, namely, 0.5. The most likely number of successes is 2 (with the average number of successes, or mean of the distribution, also 2).

HOW TO COMPUTE PROBABILITIES FOR A BINOMIAL DISTRIBUTION

Given a binomial distribution with p = probability of success and n = the number of trials, the likelihood of observing *exactly* k successes out of n trials (and, consequently, $n - k$ failures) is given by

$$\binom{n}{k} p^k (1 - p)^{n-k}, \tag{9}$$

where $\binom{n}{k}$ stands for the *binomial coefficient* and is computed by

$$\binom{n}{k} = \frac{(n)(n-1)(n-2)(n-3)\cdots\cdots(n-k+1)}{(k)(k-1)(k-2)(k-3)\cdots\cdots(1)}$$

$$= \frac{n!}{k!(n-k)!}. \tag{10}$$

The symbol $n!$ is read "n factorial" and stands for the product of all the integers between 1 and n, inclusive. For this to make sense, of course, n must be an integer 1 or greater. If $n = 0$, we define $0! = 1$. Many advanced calculators have a factorial key. If your calculator has one, you can calculate the binomial coefficient easily. If your calculator doesn't have the key, don't worry, the calculation isn't that difficult.

COMPUTING THE BINOMIAL COEFFICIENT IF YOUR CALCULATOR HAS A FACTORIAL KEY

Calculate $\binom{8}{2}$. By formula (10),

$$\binom{8}{2} = \frac{8!}{2!6!}.$$

You calculate 8! by pressing ⃞ 8 ⃞ , then ⃞ ! ⃞ . Your calculator should be displaying   *40320* , which is $1 \times 2 \times 3 \times 4 \times 5 \times 6 \times 7 \times 8$. Now press ⃞ ÷ ⃞ , ⃞ 2 ⃞ , ⃞ ! ⃞ , ⃞ = ⃞ , and you have divided 8! by 2!. Your calculator should now be displaying   *20 160* . Now again press ⃞ ÷ ⃞ , then ⃞ 6 ⃞ , ⃞ ! ⃞ , ⃞ = ⃞ , and you have divided by 6!. Your calculator should be displaying   *28* , the answer. Note that the easiest way to divide one number (8!) by the product of two numbers (2! and 6!) is to divide by the first number in the denominator and then divide *again* by the second number in the denominator.

COMPUTING THE BINOMIAL COEFFICIENT IF YOUR CALCULATOR HAS NO FACTORIAL KEY

Using formula (10), note that

$$\binom{8}{2} = \frac{(8)(7)(\cancel{6})(\cancel{5})(\cancel{4})(\cancel{3})(\cancel{2})(\cancel{1})}{(2)(1)(\cancel{6})(\cancel{5})(\cancel{4})(\cancel{3})(\cancel{2})(\cancel{1})} = \frac{(8)(7)}{(2)(1)} = 28.$$

Note that there are as many terms in the numerator as in the denominator, and that number is the same as k. By the way, if $k = 0$, then take the binomial coefficient to be 1: $\binom{n}{0} = 1$, for every n.

COMPUTING THE PROBABILITY

You can now proceed to find the probability of 2 successes out of 8 trials. Assuming $p = 0.3$, use formula (9) to get

$$\binom{8}{2} (0.3)^2 (0.7)^6.$$

Since $\binom{8}{2} = 28$, you can easily compute

$$(28)(0.3)^2(0.7)^6 = 0.2964755.$$

Now try calculating the probability of 2 successes out of 8 trials with $p = 0.4$. Write your answer here: _____.
Did you get 0.2090189? If not, here's how to make the calculation:

$$\binom{8}{2}(0.4)^2(0.6)^6 = (28)(0.4)^2(0.6)^6 = 0.2090189.$$

This is how the entire table of binomial probabilities is calculated.

MORE INFORMATION ON THE BINOMIAL DISTRIBUTION

We've already seen that there is one binomial distribution for each value of n and each probability p of success. The *mean* of this binomial distribution is given by np (the product of n and p). This means that if we repeatedly run a binomial experiment of n trials with fixed p, we expect *on average* that we will observe np successes. The *variance* of the binomial distribution is given by $np(1 - p)$. This will be of use to us later.

EXERCISE 12   Calculate the probability, or likelihood, of observing 5 successes out of 7 binomial trials, if the probability of success $= 0.7$. (*Hint*: $n = 7$, $k = 5$, $p = 0.7$.)

_____

EXERCISE 13   Calculate the complete binomial distribution with $n = 5$, $p = 0.8$. (*Hint*: You're going to be calculating six values: $k = 0, 1, 2, 3, 4, 5$. Make sure your six values add up to approximately 1.)

_____   _____   _____

_____   _____   _____

THE NORMAL APPROXIMATION TO THE BINOMIAL DISTRIBUTION

Later on, in Chapter 4, we're going to need to know the values of parts of certain binomial distributions. For example, we will need to know, for $p = 0.5$, the probability of finding between 45 and 60 successes out of 100 trials. (Don't worry now about *why* we might need such a thing; that part will become clear later.) We could use formula (9) to compute the probability of 45 successes, the probability of 46 successes, and so on. We would have to use formula (9) 16 times, and then add up the 16 values we got in order to determine this probability. [But if you think about it for a moment or two, you will see that applying formula (9) when we have large values for n and k will present some difficulties!]

Fortunately for us, there is a simpler way of computing this value, and most other similar binomial distribution problems. This involves the fact that if n is large enough, and if p is not too close to 0 or to 1, the binomial distribution is approximately normal. Specifically, we require that $np > 5$ and $n(1 - p) > 5$. In this example, $np = (100)(0.5) = 50 > 5$, and similarly for $n(1 - p)$. Consequently, to compute the probability of between 45 and 60 successes (inclusive) out of 100 trials, $p = 0.5$, we need only study the normal distribution, which (as we have said) approximates this binomial distribution, and calculate the area between two (z-score) values in the normal distribution. (See Figure 2-35.)

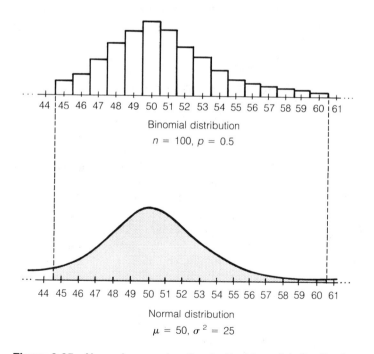

Figure 2-35 Normal approximation to the binomial distribution.

Notice in Figure 2-35 that the area we wish to determine in the binomial distribution is the sum of the areas of the bars over the values 45, 46, 47, . . . , 60. And notice *carefully* that in the normal approximation below the binomial the area begins at 44.5, which is where the bar over the value 45 begins. (The bar over any number x extends from $x - 0.5$ to $x + 0.5$.) Notice also that the area extends to 60.5, which is where the bar over the value 60 ends. This is a very important concept, called the *continuity correction to the normal approximation*. It means that you will always be adding or subtracting 0.5 from the number of successes or failures in the binomial before you compute your z-score. Further examples will make it clear just when you add 0.5 and when you subtract 0.5.

Now that we've observed all this, it's time to make our calculation: we wish the probability of between 45 and 60 successes, inclusive, and we have noted that we will compute the area in the normal distribution between 44.5 and 60.5, after making our continuity correction.

From previous remarks, we know that the mean of the binomial is

$$np = (100)(0.5) = 50$$

and the variance is

$$np(1 - p) = (100)(0.5)(0.5) = 25$$

so $\sigma = 5$. Now we compute the z-score of 44.5:

$$z = \frac{44.5 - 50}{5} = -1.1,$$

and we compute the z-score of 60.5:

$$z = \frac{60.5 - 50}{5} = +2.1.$$

Thus, the area between 44.5 and 60.5 under the normal distribution of mean 50 and variance 25 is the area between $z = -1.1$ and $z = +2.1$ in the standard normal distribution. From Table 1, this area is $0.3643 + 0.4821 = 0.8464$. So, the probability of observing between 45 and 60 successes, inclusive, out of 100 trials, with $p = 0.5$, is approximately 0.8464. [The exact value of the binomial, using formula (9), and adding up all the values, is 0.8467731.]

To deal again with the issue of when you add 0.5 and when you subtract 0.5, remember that the bar extends from 0.5 less than the value to 0.5 more than the value. If the bar is to be included, as in calculating "70 or more successes," since you will be including the value 70, and its bar begins at 69.5, you'll begin your area at 69.5. If the bar is not to be included, as in calculating "more than 70 successes," you'll begin at 70.5, since "more than 70 successes" means "71 or more successes." When in doubt, draw yourself a picture.

EXERCISE 14   Use the normal approximation to the binomial distribution to compute the probability

a. of flipping a *fair coin* ($p = 0.5$) 100 times and getting *more than* 55 heads

_____

_____

b. of getting 40 *or more* heads. Remember the continuity correction. (*Hint*: Draw a picture.)

_____

_____

EXERCISE 15   Do as requested in Exercise 14, but this time the probability of heads is 0.40.

a. _____

_____

b. _____

_____

PROBLEM SET A   1. Compute the variance of the following population of 50 corporation presidents' ages:

58	59	63	61	62	47	56	62	63	51
69	58	67	68	59	67	60	60	54	63
54	67	66	57	65	57	55	58	65	51
59	70	63	59	74	71	58	51	43	44
61	59	61	70	61	63	54	60	60	71

2. Compute the variance of the following population of 26 daily expenses:

44.98	66.56	49.28	68.03	54.95	58.49	54.68	51.37	39.06
50.05	47.57	61.20	72.60	77.11	41.24	47.97	71.35	42.91
65.03	57.54	43.54	69.17	45.39	38.89	59.97	36.43	

3. Compute the variance of the following population of 45 male models' heights, in inches:

71	72	70	71	72	72	70	70	71	70
73	72	70	72	70	74	72	72	73	71
71	72	71	73	72	71	72	71	74	71
71	71	74	73	72	71	72	70	71	71
73	70	70	72	72					

4. Compute the variance of the following population of 15 students' attention spans, in minutes:

10.9	10.1	9.2	13.6	9.9	9.7	13.8	13.0	5.0	6.6
8.6	14.0	8.5	9.9	10.1					

5. Compute the variance of the following population of 30 automobile gasoline mileages:

21.9	38.6	34.7	31.2	17.2	18.6	28.4	30.3	35.7	21.9
23.1	38.8	38.0	37.3	16.2	22.7	28.3	22.4	24.0	15.6
31.5	22.7	24.4	16.9	26.6	21.8	30.6	19.6	35.8	23.0

6. Compute the variance of the following population of 35 television repair bills:

60.33	55.82	55.99	165.58	96.48	154.74	137.06
108.82	67.42	79.94	87.44	159.72	88.50	175.32
165.41	153.53	104.00	178.08	49.35	65.01	119.59
121.20	130.35	130.98	154.85	155.94	124.60	52.77
75.76	94.34	126.01	158.76	107.22	111.72	98.59

7. Compute the variance of the following population of 15 rats' weight gains, in grams:

8.8	28.3	9.6	9.6	42.0	43.3	5.3	19.6	34.2	13.2
4.2	45.9	5.9	12.6	10.7					

8. Compute the variance of the following population of 32 executive salaries:

128,000	145,000	78,000	136,000	60,000	72,000	91,000
217,000	46,000	191,000	85,000	127,000	139,000	134,000
182,000	104,000	173,000	177,000	197,000	213,000	149,000
112,000	133,000	145,000	101,000	155,000	167,000	143,000
213,000	88,000	149,000	173,000			

9. Compute the variance of the following population of 60 pocketsful of change, in cents:

7	18	0	89	4	24	6	21	21	15
19	45	55	71	32	29	54	62	33	38
11	79	69	51	86	76	60	63	28	72
0	32	57	18	85	27	63	48	58	37
86	25	15	74	5	21	62	26	68	1
87	81	84	71	1	78	83	30	8	69

10. Compute the variance of the following population of 30 flying times (in minutes) to cross the United States:

334	317	290	289	285	264	294	280	324	301
306	316	286	328	358	276	358	308	325	354
289	281	253	268	328	270	268	318	289	332

11. Compute the variance of the following population of 33 measurements of galvanic skin response, in volts:

0.0213	0.0182	0.0201	0.0188	0.0204	0.0181	0.0190
0.0198	0.0214	0.0204	0.0208	0.0206	0.0210	0.0206
0.0200	0.0215	0.0206	0.0199	0.0206	0.0210	0.0190
0.0192	0.0202	0.0218	0.0181	0.0217	0.0180	0.0180
0.0182	0.0201	0.0197	0.0190	0.0214		

PROBLEM SET B

1. The following is a sample of size 50 from a population of corporation presidents' ages. Compute the sample variance.

58	59	63	61	62	47	56	62	63	51
69	58	67	68	59	67	60	60	54	63
54	67	66	57	65	57	55	58	65	51
59	70	63	59	74	71	58	51	43	44
61	59	61	70	61	63	54	60	60	71

2. The following is a sample of size 26 from a population of daily expenses. Compute the sample variance.

44.98	66.56	49.28	68.03	54.95	58.49	54.68	51.37	39.06
50.05	47.57	61.20	72.60	77.11	41.24	47.97	71.35	42.91
65.03	57.54	43.54	69.17	45.39	38.89	59.97	36.43	

3. The following is a sample of size 45 from a population of male models' heights, in inches. Compute the sample variance.

71	72	70	71	72	72	70	70	71	70
73	72	70	72	70	74	72	72	73	71
71	72	71	73	72	71	72	71	74	71
71	71	74	73	72	71	72	70	71	71
73	70	70	72	72					

4. The following is a sample of size 15 from a population of students' attention spans, in minutes. Compute the sample variance.

10.9	10.1	9.2	13.6	9.9	9.7	13.8	13.0	5.0	6.6
8.6	14.0	8.5	9.9	10.1					

5. The following is a sample of size 30 from a population of automobile gasoline mileages. Compute the sample variance.

21.9	38.6	34.7	31.2	17.2	18.6	28.4	30.3	35.7	21.9
23.1	38.8	38.0	37.3	16.2	22.7	28.3	22.4	24.0	15.6
31.5	22.7	24.4	16.9	26.6	21.8	30.6	19.6	35.8	23.0

6. The following is a sample of size 35 from a population of television repair bills. Compute the sample variance.

60.33	55.82	55.99	165.58	96.48	154.74	137.06
108.82	67.42	79.94	87.44	159.72	88.50	175.32
165.41	153.53	104.00	178.08	49.35	65.01	119.59
121.20	130.35	130.98	154.85	155.94	124.60	52.77
75.76	94.34	126.01	158.76	107.22	111.72	98.59

7. The following is a sample of size 15 from a population of rats' weight gains, in grams. Compute the sample variance.

8.8	28.3	9.6	9.6	42.0	43.3	5.3	19.6	34.2	13.2
4.2	45.9	5.9	12.6	10.7					

8. The following is a sample of size 32 from a population of executive salaries. Compute the sample variance.

128,000	145,000	78,000	136,000	60,000	72,000	91,000
217,000	46,000	191,000	85,000	127,000	139,000	134,000
182,000	104,000	173,000	177,000	197,000	213,000	149,000
112,000	133,000	145,000	101,000	155,000	167,000	143,000
213,000	88,000	149,000	173,000			

9. The following is a sample of size 60 from a population of pocketsful of change, in cents. Compute the sample variance.

7	18	0	89	4	24	6	21	21	15
19	45	55	71	32	29	54	62	33	38
11	79	69	51	86	76	60	63	28	72
0	32	57	18	85	27	63	48	58	37
86	25	15	74	5	21	62	26	68	1
87	81	84	71	1	78	83	30	8	69

10. The following is a sample of size 30 from a population of flying times (in minutes) to cross the United States. Compute the sample variance.

334	317 '	290	289	285	264	294	280	324	301
306	316	286	328	358	276	358	308	325	354
289	281	253	268	328	270	268	318	289	332

11. The following is a sample of size 33 from a population of measurements of galvanic skin response, in volts. Compute the sample variance.

0.0213	0.0182	0.0201	0.0188	0.0204	0.0181	0.0190
0.0198	0.0214	0.0204	0.0208	0.0206	0.0210	0.0206
0.0200	0.0215	0.0206	0.0199	0.0206	0.0210	0.0190
0.0192	0.0202	0.0218	0.0181	0.0217	0.0180	0.0180
0.0182	0.0201	0.0197	0.0190	0.0214		

PROBLEM SET C

1. The area between $z = -1.25$ and $z = +1.18$ is _____.

2. The area between $z = -0.84$ and $z = +1.72$ is _____.

3. The area between $z = -2.70$ and $z = -1.07$ is _____.

4. The area between $z = -1.51$ and $z = -0.41$ is _____.

5. The area between $z = -0.40$ and $z = -0.17$ is _____.

6. The area between $z = -2.05$ and $z = +2.27$ is _____.

7. The area between $z = -0.75$ and $z = +1.92$ is _____.

8. The area between $z = -0.01$ and $z = +2.75$ is _____.

9. The area between $z = +0.18$ and $z = +0.86$ is _____.

10. The area between $z = -1.19$ and $z = +1.48$ is _____.

11. The area between $z = -2.58$ and $z = -0.08$ is _____.

12. The area between $z = -2.02$ and $z = +2.03$ is _____.

13. The area between $z = -2.52$ and $z = +1.40$ is _____.

14. The area between $z = +0.02$ and $z = +1.86$ is _____.

15. The area between $z = -1.77$ and $z = -0.39$ is _____.

16. The area between $z = -1.11$ and $z = +1.09$ is _____.

17. The area between $z = -1.85$ and $z = +1.42$ is _____.

18. The area between $z = -1.80$ and $z = +0.65$ is _____.

19. The area between $z = -0.99$ and $z = +2.36$ is _____.

20. The area between $z = -1.81$ and $z = -1.41$ is _____.

21. The area between $z = -2.80$ and $z = +1.14$ is _____.

22. The area between $z = -0.09$ and $z = +2.20$ is _____.

23. The area between $z = +1.52$ and $z = +2.64$ is _____.

24. The area between $z = -2.06$ and $z = +0.66$ is _____.

25. The area between $z = +1.16$ and $z = +1.78$ is _____.

PROBLEM SET D

1. A normal distribution is known to have mean 40 and standard deviation 10. Convert the following numbers to z-scores:

    30.00   15.30   61.50   47.20   43.90

2. A normal distribution is known to have mean 300 and standard deviation 50. Convert the following numbers to z-scores:

    369.50   198.00   233.50   197.50   299.50

3. A normal distribution is known to have mean -50 and standard deviation 5. Convert the following numbers to z-scores:

    -50.90 -41.85 -45.45 -54.95 -57.20

4. A normal distribution is known to have mean 1000 and standard deviation 400. Convert the following numbers to z-scores:

    1596.00   2020.00   196.00 -24.00 1356.00

5. A normal distribution is known to have mean 80 and standard deviation 25. Convert the following numbers to z-scores:

$$74.50 \quad 19.00 \quad 141.00 \quad 85.75 \quad 136.00$$

6. A normal distribution is known to have mean 0 and standard deviation 1. Convert the following numbers to z-scores:

$$2.16 \quad 2.13 \quad -1.84 \quad 0.92 \quad 0.24$$

7. A normal distribution is known to have mean 0 and standard deviation 100. Convert the following numbers to z-scores:

$$-296.00 \quad -51.00 \quad -179.00 \quad -1.00 \quad 47.00$$

8. A normal distribution is known to have mean 100 and standard deviation 16. Convert the following numbers to z-scores:

$$111.68 \quad 60.16 \quad 108.96 \quad 80.96 \quad 115.68$$

PROBLEM SET E

1. A normal distribution is known to have mean 40 and standard deviation 10. Find the area between 14.10 and 28.80.

2. A normal distribution is known to have mean 300 and standard deviation 50. Find the area between 160.00 and 196.00.

3. A normal distribution is known to have mean -50 and standard deviation 5. Find the area between -52.00 and -41.10.

4. A normal distribution is known to have mean 1000 and standard deviation 400. Find the area between 2080.00 and 2088.00.

5. A normal distribution is known to have mean 80 and standard deviation 25. Find the area between 68.75 and 153.50.

6. A normal distribution is known to have mean 0 and standard deviation 1. Find the area between -0.46 and 1.24.

7. A normal distribution is known to have mean 0 and standard deviation 100. Find the area between -247.00 and -25.00.

8. A normal distribution is known to have mean 100 and standard deviation 16. Find the area between 98.40 and 112.00.

PROBLEM SET F

1. Calculate the probability, or likelihood, of observing 2 successes out of 8 binomial trials, if the probability of success = 0.30.

2. Calculate the probability, or likelihood, of observing 2 successes out of 8 binomial trials, if the probability of success = 0.40.

3. Calculate the probability, or likelihood, of observing 3 successes out of 10 binomial trials, if the probability of success = 0.30.

4. Calculate the probability, or likelihood, of observing 5 successes out of 7 binomial trials, if the probability of success = 0.50.

5. Calculate the probability, or likelihood, of observing 3 successes out of 5 binomial trials, if the probability of success = 0.60.

6. Calculate the probability, or likelihood, of observing 8 successes out of 10 binomial trials, if the probability of success = 0.90.

7. Calculate the probability, or likelihood, of observing 12 successes out of 15 binomial trials, if the probability of success = 0.80.

8. Calculate the probability, or likelihood, of observing 1 success out of 20 binomial trials, if the probability of success = 0.05.

9. Calculate the probability, or likelihood, of observing 2 successes out of 4 binomial trials, if the probability of success = 0.50.

10. Calculate the probability, or likelihood, of observing 5 successes out of 10 binomial trials, if the probability of success = 0.50.

PROBLEM SET G

1. A fair coin ($p = 0.5$) is tossed 100 times. Use the normal approximation to the binomial distribution to compute the probability

   a. of observing more than 40 heads
   b. of observing 40 or more heads

2. Same as Problem 1, but now the probability of heads is 0.40. Compute the probability

   a. of observing more than 40 heads
   b. of observing 40 or more heads

3. A fair coin ($p = 0.5$) is tossed 100 times. Use the normal approximation to the binomial distribution to compute the probability

   a. of observing more than 60 heads
   b. of observing 60 or more heads

4. Same as Problem 3, but now the probability of heads is 0.55. Compute the probability

   a. of observing more than 60 heads
   b. of observing 60 or more heads

5. A fair coin ($p = 0.5$) is tossed 60 times. Use the normal approximation to the binomial distribution to compute the probability

   a. of observing more than 30 heads
   b. of observing 30 or more heads

6. Same as Problem 5, but now the probability of heads is 0.40. Compute the probability

   a. of observing more than 30 heads
   b. of observing 30 or more heads

7. A fair coin ($p = 0.5$) is tossed 200 times. Use the normal approximation to the binomial distribution to compute the probability

   a. of observing more than 120 heads
   b. of observing 120 or more heads

8. Same as Problem 7, but now the probability of heads is 0.61. Compute the probability

   a. of observing more than 120 heads
   b. of observing 120 or more heads

9. A fair coin ($p = 0.5$) is tossed 1000 times. Use the normal approximation to the binomial distribution to compute the probability

   a. of observing more than 500 heads
   b. of observing 500 or more heads

10. Same as Problem 9, but now the probability of heads is 0.40. Compute the probability

    a. of observing more than 500 heads
    b. of observing 500 or more heads

11. A fair coin ($p = 0.5$) is tossed 1000 times. Use the normal approximation to the binomial distribution to compute the probability

    a. of observing more than 600 heads
    b. of observing 600 or more heads

12. Same as Problem 11, but now the probability of heads is 0.70. Compute the probability

    a. of observing more than 600 heads
    b. of observing 600 or more heads

CHAPTER 2 QUIZ

1. Variation is a measure of _____
   of a population.

2. Standard deviation is the _____
   of variance.

3. The sample variance is an _____

   of the population _____ .

4. Distribution is another name for the _____
   of a statistic.

5. For a continuous distribution, the proportion of the population or distribution that lies between the values 2 and 3 is given by the _____

_____ .

6. A large number, say 150 or more, of IQs, of repeated measurements of a physical constant, or of weights of oak leaves are all examples of _____ distributions.

7. The area between -1.96 and $+1.96$, under the normal distribution, is ____% of the total area.

8. "Independent trials process" is another name for _____ .

9. $\binom{n}{k}$ is called a _____ .

ANSWERS TO CHAPTER 2 EXERCISES

1.

x_i	$x_i - \mu$	$(x_i - \mu)^2$
0	$0 - 5 = -5$	$(-5)^2 = 25$
5	$5 - 5 = 0$	0
5	$5 - 5 = 0$	0
5	$5 - 5 = 0$	0
5	$5 - 5 = 0$	0
5	$5 - 5 = 0$	0
5	$5 - 5 = 0$	0
5	$5 - 5 = 0$	0
5	$5 - 5 = 0$	0
10	$10 - 5 = 5$	25

$$\Sigma(x_i - \mu)^2 = 50$$

$$\text{var}(E) = \sigma_E^2 = \frac{\Sigma(x_i - \mu)^2}{10} = \frac{50}{10} = 5$$

2. 6.44

3. 5.56

4. 5.27, 2.54, 2.36, respectively. The three answers are obtained by taking the square root of each of the numbers 27.78, 6.44, and 5.56, which were the sample variances.

5. a. and b. D: $\sigma_{n-1} = 2.5377155$; variance $= 6.44$
              E: $\sigma_{n-1} = 2.3579652$; variance $= 5.56$

6. a. and b. $\sigma_D^2 = 5.8$; $\sigma_E^2 = 5$

7. $0, 0.5, 1.5, 2.5, -0.5, -1$

8. $1, 2, 3, 0, 0.5, 0.25, -1$

9. (60, 140)

   Standard deviation $= \sigma = \sqrt{\text{variance}} = \sqrt{400} = 20$;

   $2\sigma = 40$; $100 \pm 40 = (60, 140)$

10. a. $0.4788 - 0.4406 = 0.0382$

    b. The same value, as the distribution is symmetric and this is the same area, but on the left side of the normal distribution.

    c. $0.4406 + 0.4788 = 0.9194$. We *add* values because one z-score is negative and the other is positive.

11. $\dfrac{70.3 - 50}{10} = \dfrac{20.3}{10} = 2.03$; $\dfrac{65.0 - 50}{10} = \dfrac{15}{10} = 1.5$

    The area between z-scores of 1.5 and 2.03 is $0.4788 - 0.4332 = 0.0456$.

12. $\dbinom{7}{5}(0.7)^5(0.3)^2 = (21)(0.7)^5(0.3)^2 = 0.3176523$

13. 0.00032; 0.0064; 0.0512; 0.2048; 0.4096; 0.32768

14. a. "*More than 55*" means 56, 57, 58, 59, 60, . . . , 100. The histogram looks like:

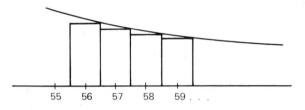

So the z-score is to be computed, after the continuity correction, at $55\frac{1}{2}$ (because that's where the bar for 56 starts).

$$n = 100; \quad p = \tfrac{1}{2}; \quad np = 50; \quad z = \frac{55\frac{1}{2} - 50}{\sqrt{(100)(\frac{1}{2})(\frac{1}{2})}} = 1.1$$

From Table 1, $\text{Prob}(z \geq 1.1) = 0.1357$:

$$\text{Prob}(z \geq 1.1) = \text{Probability that } z \text{ is greater than } 1.1$$
$$= 0.5000 - \text{Prob}(0 \leq z \leq 1.1),$$

which, from Table 1, is

$$0.5000 - 0.3643 = 0.1357.$$

b. For 40 or more heads, we compute from $39\frac{1}{2}$.

$$z = \frac{39\frac{1}{2} - 50}{\sqrt{(100)(\frac{1}{2})(\frac{1}{2})}} = -2.1$$

$$\text{Prob}(z \geq -2.1) = 0.50 + 0.4821 = 0.9821$$

15. Same as Exercise 14, but $p = 0.4$; $np = 40$.

a.
$$z = \frac{55\frac{1}{2} - 40}{\sqrt{100(0.4)(0.6)}} = 3.16$$

Table 1 gives values for z-scores only to $z = 3.09$.
$\text{Prob}(z \geq 3.09) = 0.5000 - 0.4990 = 0.001$, so $\text{Prob}(z \geq 3.16) < 0.001$.

b.
$$z = \frac{39\frac{1}{2} - 40}{\sqrt{100(0.4)(0.6)}} = -0.10$$

$$\text{Prob}(z \geq -0.10) = \text{Prob}(-0.10 \leq z \leq 0) + \text{Prob}(z \geq 0)$$

$$= 0.0398 + 0.5000$$

$$= 0.5398$$

ANSWERS TO CHAPTER 2 QUIZ

1. spread, variability

2. square root

3. estimator; variance

4. histogram

5. area under the curve between 2 and 3

6. normal

7. 95%. Look up 1.96 in Table 1, and get 0.4750. Then double it.

8. binomial process, or binomial trials

9. binomial coefficient

Statistical Inference

Terms we'll be learning about in this chapter

confidence intervals

estimation

proportions

quality control

3-1 ESTIMATION OF THE MEAN

THE PROBLEM

Fred's Tire Shop has an inventory of 1000 tires that Fred would like to dispose of in a single fleet sale. Now suppose you and I were prospective buyers, owners of a courier service that operates several hundred cars. We've agreed with Fred on a sales price for the tires based on the average tire mileage lifetime of the tires: $1 per thousand miles per tire. That means that if the tires will last an average of 40,000 miles, we will pay $40 per tire, or $40,000. If the tires will last an average of 30,000 miles, we will pay $30 per tire, or $30,000 for 1000 tires. So you can see that the average lifetime of the 1000 tires is of considerable interest, both to us and to Fred.

One way to determine the mean tire lifetime for all 1000 tires is to *test* each tire by putting it on a machine that will simulate the wearing of tires, and see how long it lasts. But if we test a tire, we use it up (called **destructive testing**, as with bullets and light bulbs), and so if we tested every tire, we'd have none left to buy. For this reason we are forced to study a sample of tires from the complete population of 1000 tires and from that sample infer what the mean of the population is.

SAMPLING

In order to make the sale, Fred has agreed to provide us with a sample of 30 tires, chosen at random from the 1000, to be tested on a tire mileage machine. The tires are selected and tested, and the 30 tire mileages are as follows:

38,900	38,100	38,200	39,600	41,400	37,200
37,000	35,500	41,500	41,000	37,800	37,200
37,500	36,300	35,700	35,800	38,200	38,100
37,900	37,500	36,500	39,400	39,600	41,000
35,000	38,200	36,800	38,100	38,100	36,900

$$(n = 30)$$

For this sample, $n = 30$ (the sample size is 30) and the mean $\bar{x} = 38,000$. Now given that the sample mean is 38,000, just what can we say about the population mean? Put another way, when we draw a sample of size 30 from a population, how often is the mean of the sample close to the mean of the population? *This question, and its answer, form the basis for statistical inference.*

To discover what information the sample of 30 numbers gives us about the population of 1000 numbers—specifically about the population mean—we must first ask where the sample of size 30 comes from. It is only one of all the possible samples of size 30 from a population of size 1000. Another possible sample of size 30 would be the 30 largest tire mileages. And still another possible sample would be the 30

smallest tire mileages. Actually, there is an enormous number of possible samples of size 30. There are

$$\binom{1000}{30} = 2{,}429{,}608{,}192{,}173{,}745{,}103{,}270{,}389{,}838{,}576{,}750{,}719{,}302{,}222{,}606{,}198{,}631{,}438{,}800$$

possible samples. This number is so large that it is usually rounded off and written in scientific notation: 2.4×10^{57}. From this very large population of samples of size 30 we draw one member, that is, one sample of size 30. We are going to make an inference about the mean of the population of mileages of 1000 tires based on this one sample, together with consideration of all those 2.4×10^{57} possible samples.

As I said above, of all the possible samples of size 30, it *could* be that the one we draw will consist of the 30 largest values, or the 30 smallest values. This is possible but, fortunately, not very likely. It's much more likely that some of the numbers will be larger than the population mean and some will be smaller, so that the mean of the sample will lie somewhat close to the population mean (mean of 1000 tire mileages). Refer back to Figure 2-9. What we are saying is that the first part of Figure 2-9 is *much more likely* to occur than the second part. This is a key idea in our estimation.

The closer the sample mean falls to the population mean, the better our estimate will be. And the more often this happens, the better. This issue of "how close . . . how often" is what we need to understand next.

CONFIDENCE INTERVALS

Let's suppose first (and this *isn't* true) that the mean of the sample always lies within 100 (miles) of the mean of the population. Then we can reason like this: since the sample mean is within 100 (miles) of the population mean, the population mean is within 100 of the sample mean. ("If A is close to B, then B is close to A"; this point is often misunderstood. If you don't see what the issue is, you're missing the point. Go back and read it over.) So if we go out ± 100 miles from the sample mean $\bar{x}$, we will always find the population mean within that interval $\bar{x} \pm 100$. In our case, $\bar{x} = 38,000$, so the interval will be (37,900, 38,100). That is, the population mean lies between 37,900 and 38,100. Thus, we have estimated the population mean from information about the sample mean.

But in real life, statements about "how close" the sample mean is to the population mean are never going to be true *all the time*. There will always be the one sample that consists of the 30 largest values from the population, or the 30 smallest values, and when we get one of these samples, our sample mean is going to be very far from the population mean. Fortunately, this happens only rarely (see Figure 3-1).

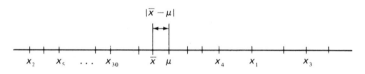

Some of the x_js are larger than μ, and some of the x_js are smaller than μ. $\bar{x}$ is close to μ. $|\bar{x} - \mu|$ is small. This happens most of the time.

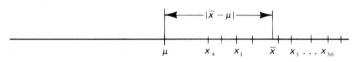

All (or most) of the x_js fall to one side of μ. $\bar{x}$ is far from μ. $|\bar{x} - \mu|$ is large. This happens only rarely.

Figure 3-1

While a statement like "the sample mean *always* lies within 100 of the population mean" is not feasible, a statement like "the sample mean lies within 100 of the population mean 90% of the time" is possible. (This statement isn't true either, but its form is correct. Let's understand it.) Just what does "90% of the time" mean? We now consider *all* the 2.4×10^{57} samples of size 30. We may be able to determine that 90% of these samples have the property that their sample mean lies within 100 of the population mean. Then, since each sample of size 30 is as likely as any other sample of size 30 (since the tires were drawn randomly), 90% of the time that we draw out a sample of size 30, the mean of that sample will lie within 100 of the population mean; and therefore, if we draw out a sample of size 30 (and it is one of those 90%) and if we form an interval ±100 about the sample mean, the population mean will lie within that interval. (It is also true that 10% of the time we do this the sample mean will be one of those that do *not* lie within 100 of the population mean. Then if we form an interval ±100 about the sample mean, $\bar{x} \pm 100$, the population mean will *not* lie in the interval.) Whether we have one of the samples for which the sample mean lies within 100 of the population mean (true for 90% of all of the 2.4×10^{57} samples) or one of the samples for which the sample mean does not lie within 100 of the population mean (true for 10% of all the samples), we do not know. But we do know that if we form an interval

$$\bar{x} \pm 100 = 38{,}000 \pm 100 = (37{,}900, \ 38{,}100)$$

about the sample mean and state that the population mean lies within that interval, we have a 90% chance of being correct and a 10% chance of being wrong. Whether we are right or wrong depends on the "luck of the draw." This is a statement of **confidence interval type**: "the population mean lies in the interval $(37{,}900, \ 38{,}100)$," or "$37{,}900 \leq \mu \leq 38{,}100$." It is a statement that will be true 90% of the time. The interval $(37{,}900, \ 38{,}100)$ is called a **90% confidence interval** for μ.

An issue that bothers some readers goes like this: "How can I make a statement that is *either right or wrong* and call it correct 90% of the time? Either it's right or it's wrong!" Well, that's the same issue I face when I flip a fair coin and without looking say, "It's heads." I am making a statement that would be correct 50% of the time if I were to repeat this procedure a large number of times. Now, of course, I may not flip the coin a large number of times. I may just do it once. Nonetheless, I can say, "It's heads," and be making a statement at the 50% confidence level.

Now, where did I get the information that 90% of the 2.4×10^{57} samples of size 30 will have their sample means lie within ±100 of the population mean? Actually, I made that up. I said earlier that this statement wasn't true, but that the *form* of it was correct; I did that just so we could study and understand statements of confidence interval type. Now let's talk about just exactly "how close . . . how often."

THE DISTRIBUTION OF SAMPLE MEANS

We have seen how to construct statements of confidence interval type if we know how close the sample means fall to the population mean how often. The question of "how close . . . how often" is answered if we know the histogram that describes the distribution of *all* 2.4×10^{57} sample means for all 2.4×10^{57} samples of size 30. These are all the possible values for the statistic $\bar{x}_{30}$, all the sample means of samples of size 30. For now we're going to assume that the sample size is always 30; we will occasionally refer to the distribution of sample means, or DSM, without mentioning the number 30. Shortly, we'll consider the case for DSMs of other sample sizes. When we say DSM, we mean, of course, the histogram (see Figure 3-2). This distribution will give us the information of "how close . . . how often," which will allow us to form statements of confidence interval type. It will allow us to say what a 90% confidence interval is, a 95% confidence interval, a 99% confidence interval, and so on.

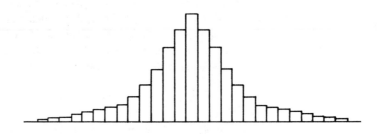

Figure 3-2   Distribution of $\bar{x}_{30}$, sample means of samples of size 30 (DSM).

There are three things we will need to know about this distribution:

1. The mean

2. The variance

3. The shape

1. The Mean.   *The mean of the DSM is the same as the mean of the population.* To see this, we need to ask, What is the mean of all the 2.4×10^{57} sample means? This question is easy to answer when we consider that each sample is a sample of 30 chosen from the population of 1000 mileages and that each of the 1000 mileages occurs in exactly as many of the 2.4×10^{57} samples as every other one of the 1000 mileages. Then the mean of the 2.4×10^{57} sample means must be the same as the mean of the 1000 mileages, since each of the mileages occurs as often in the samples as every other. But the mean of the 1000 mileages is the population mean, the thing we are trying to estimate. Therefore, the mean of the DSM is the population mean μ.

2. The Variance.   It is a mathematical fact that the variance of the DSM for sample size 30 is

$$\sigma_{\bar{x}}^2 = \sigma_{\text{DSM}}^2 = \frac{\text{Population variance}}{30} = \frac{\sigma_{\text{pop}}^2}{30} \, . \quad * \qquad (1)$$

*Note: σ_{DSM}^2 is usually written $\sigma_{\bar{x}}^2$, since what we are discussing is the distribution of sample means, $\bar{x}$.

The reason we measure *squared* deviations rather than, say, absolute deviations, or cubed deviations, is that when we choose our definition of variance to be squared deviations, it makes formula (1) true. There would be no such formula if variance were defined differently.

3. The Shape.   For sample sizes 30 or more taken from most populations (including the tire lifetime population), the DSM will be approximately *normal*. This remarkable theorem is called the *central limit theorem*. It is remarkable because no matter what the distribution of the original population looks like, the distribution of the sample means (DSM) from that population tends to look more and more like a bell-shaped curve, the larger the sample size. And for *most* populations, sample size 30 is large enough that the DSM will be pretty nearly normal (see Figure 3-3).

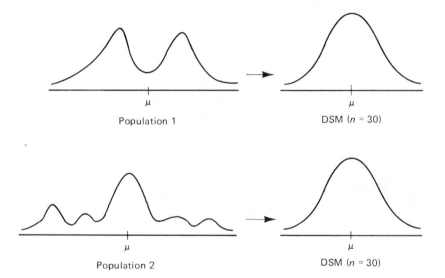

Figure 3-3

So now we have answered our three questions about the DSM:

1. The mean of the DSM is the population mean (unknown) μ.

2. The variance of the DSM is

$$\frac{\text{Population variance}}{30} = \frac{\sigma_{\text{pop}}^2}{30}$$

Note: The 30 is the sample size. If we had sample size 50, we'd divide by 50; in general we divide by n:

$$\sigma^2_{\text{DSM}} = \frac{\sigma^2_{\text{pop}}}{n} . \tag{2}$$

3. The DSM is approximately a *normal distribution*.

Assume for the moment that the population variance is known to be 1,200,000. Then according to item 2, the variance of the DSM will be 1,200,000/30 = 40,000; thus, the *standard deviation* (which is the square root of the variance) will be 200. Now we have complete information about the DSM, which is shown in Figure 3-4. We know that it is normal, that it has standard deviation 200, and that it is centered on the population mean. We are now in a position to answer all the "how close . . . how often" questions of statistical inference.

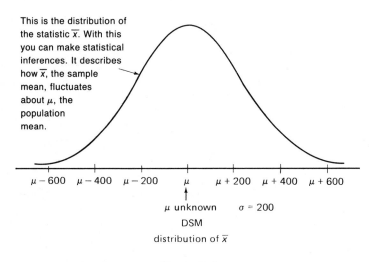

This is the distribution of the statistic $\overline{x}$. With this you can make statistical inferences. It describes how $\overline{x}$, the sample mean, fluctuates about μ, the population mean.

$\mu - 600$ $\mu - 400$ $\mu - 200$ μ $\mu + 200$ $\mu + 400$ $\mu + 600$

μ unknown $\sigma = 200$

DSM

distribution of $\overline{x}$

Figure 3-4

The region within ± 1 standard deviation (that is, 1σ) on either side of the mean of the normal distribution is about 68% of the total area of the distribution. This can be calculated from the information in Table 1. However, to simplify matters we have constructed Table 2, which is a table for confidence intervals for the normal distribution. We have printed a portion of Table 2 as Table 3-1. So in the case of the DSM in Figure 3-5, we know that 68% of the time we draw a sample of size 30 and compute its mean, that sample mean $\overline{x}$ will lie within 200 of the population mean (because the standard deviation σ of this distribution is assumed to be 200). There-

Table 3-1   Normal Table for Confidence Intervals (from Table 2)

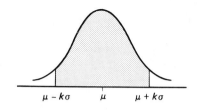

When you go out ± this many sigmas from the mean of the normal distribution (k)	You capture this percentage of the normal distribution
0	0%
1	68.3%
1.645	90.0%
1.96	95.0%
2	95.4%
2.576	99.0%
3	99.7%

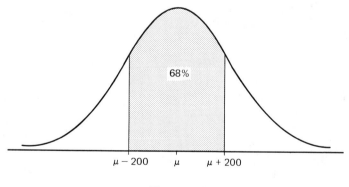

Figure 3-5

fore, the population mean (unknown) will lie within 200 of the sample mean $\bar{x}$. If we compute the sample mean $\bar{x}$ and form an interval $\bar{x} \pm 1\sigma = (\bar{x} - 200, \ \bar{x} + 200)$, the population mean (unknown) will lie inside that interval every time the sample mean lies within 200 of the population mean, that is, 68% of the time (see Figure 3-6). Therefore, $(\bar{x} - 200, \ \bar{x} + 200)$ is a 68% confidence interval. Figure 3-6 con-

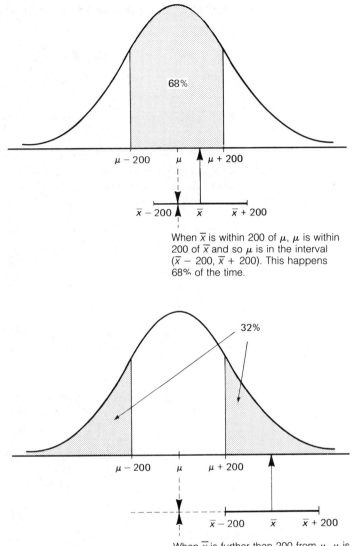

When $\bar{x}$ is within 200 of μ, μ is within 200 of $\bar{x}$ and so μ is in the interval $(\bar{x} - 200, \bar{x} + 200)$. This happens 68% of the time.

When $\bar{x}$ is further than 200 from μ, μ is further than 200 from $\bar{x}$ and so μ is not in the interval $(\bar{x} - 200, \bar{x} + 200)$. This happens 32% of the time.

Figure 3-6

tains all the essential ideas of what a 68% confidence interval means. Study it carefully until you understand it *completely*.

COMPUTING CONFIDENCE INTERVALS

We could repeat the same reasoning above and construct a confidence interval ± 2 standard deviations ($\pm 2\sigma$) from the mean instead of $\pm 1\sigma$. Then we would be forming an interval ($\bar{x} - 400$, $\bar{x} + 400$) about the sample mean, and to determine the confidence level, we would need to know *how often* the elements of a normal distribution fall within 2 standard deviations of the mean of the normal distribution. Referring to Table 2 (see Table 3-1), we see that that occurs approximately 95.4% of the time. So the confidence interval ($\bar{x} - 400$, $\bar{x} + 400$) captures the population mean 95.4% of the time. We are not limited to whole numbers of standard deviations, either: 68% of the normal distribution corresponds to ± 1 sigma; 95.4% corresponds to ± 2 sigmas; 90% corresponds to ± 1.645 sigmas. And for any percentage confidence we wish, we can determine, via Table 2, how far to go out on either side of the mean, measured by the number of standard deviations of the distribution, so that that interval will contain that desired percentage of the total distribution.

Let's review for a moment the estimation process. We draw a sample (in this case, of size 30; in general, of size n) from a population whose variance we assume we know. We compute the mean of this sample and compute the standard deviation of the DSM [via formula (2)], assuming that we know the population variance. We use the fact that we know that the DSM is approximately normal. Depending on the confidence we wish to place on our estimation, we determine from Table 2 how far on either side of the mean, measured in σ_{DSM}'s, we must go to capture that proportion of the normal distribution (68% is ± 1 sigma; 95.4% is ± 2 sigmas; 90% is ± 1.645 sigmas; and so on).

THE PROBLEM, SOLVED

Getting back to our tire mileage problem, if our sample mean is 38,000 miles and the population variance is known to be 1,200,000 (as we have assumed), then we obtain $\sigma_{DSM} = 200$, and a 68% confidence interval for μ (the population mean) is

$$\bar{x} \pm 1\sigma = 38,000 \pm 200$$

$$= (37,800, \ 38,200).$$

Similarly, a 95.4% confidence interval is

$$\bar{x} \pm 2\sigma = 38,000 \pm 2(200) = 38,000 \pm 400$$

$$= (37,600, \ 38,400),$$

and a 90% confidence interval is

$$\bar{x} \pm 1.645\sigma = 38,000 \pm (1.645)(200)$$

$$= 38,000 \pm 329$$

$$= (37,671, 38,329).$$

The wider the confidence interval, the greater the chance that the population mean will really fall inside it, and so the higher the confidence level. Conversely, the "tighter" (narrower) the confidence interval, the less confidence we can attach to it. This is the basic trade-off in inferential statistics: we can have a high degree of confidence in our estimate, but only if our estimate is not very precise. Or we can have a very precise estimate of the mean, but the likelihood that we are correct will be correspondingly less. (This is a statistical "uncertainty principle.")

There is only one thing remaining to complete our estimation. You may remember that we *assumed* that the population variance was known to be 1,200,000 for the population of tire mileages. I made that number up because it worked out to give a convenient value for σ_{DSM}. In practice, however, we won't know the population variance exactly. But for sample sizes of 30 or more, it is likely that s^2 is approximately σ_{pop}^2. So, we will simply use s^2, the sample variance defined in formula (4) of Chapter 2 (page 51), as an estimate of σ_{pop}^2, the population variance. This substitution of s^2 for σ^2 will not invalidate any of our development of confidence intervals. For sample sizes less than 30, we will need another distribution, the t-distribution, which we will discuss in Chapter 5.

Now that we know the correct, complete way to form confidence intervals, let's form the *totally correct* 90% confidence interval for the mean tire mileage:

Step 1. Compute the sample variance, $s^2 = $ _____ .
Use your calculator to do this. Your answer should be 2,960,689.7.

Step 2. Use formula (2) (page 98) to compute σ of the DSM:

$$\sigma_{DSM} = \sqrt{\frac{\sigma_{pop}^2}{30}} \cong \sqrt{\frac{s^2}{30}} = \sqrt{\frac{2,960,689.7}{30}}$$

Means "is approximately equal to"

$$= \underline{\hspace{3cm}}$$

Your answer should be 314.1.

Step 3. Form a 90% confidence interval by looking up 90% in Table 2 or Table 3-1 and seeing that we must go out $\pm 1.645\sigma_{DSM}$ from the sample mean, $\bar{x} = 38,000$.

Thus, our 90% confidence interval is

$$38,000 \pm (1.645)(314.1) = 38,000 \pm 516.77 = (37,483, \ 38,517).$$

Now we can be sure, at the 90% level of confidence, that if we pay Fred $38 per tire, at most we are overpaying about 52¢ per tire; and Fred can also be sure, at the 90% confidence level, that he is not going to be more than 52¢ underpriced (based on what the *true*, unknown mean tire mileage is).

To form a 95% confidence interval, we use the same sample mean $\bar{x} = 38,000$ and the same sample variance $s^2 = 2,960,689.7$. We look in Table 3-1, the table of confidence intervals for the normal distribution, and find that the entry for 95% is 1.96. This means we want to form the interval $\bar{x} \pm 1.96\sigma_{\text{DSM}}$. When we substitute s for σ_{pop} in formula (2), page 98, we see that our 95% confidence interval becomes

$$\bar{x} \pm 1.96 \frac{s}{\sqrt{n}} = 38,000 \pm (1.96)(314.1)$$
$$= 38,000 \pm 615.6$$
$$= (37,384, \ 38,616).$$

Notice that the 95% confidence interval is *wider* than the 90% confidence interval, which is as it should be, because to be more confident we must be less precise. (That is, if we want to capture the population mean in our interval more often, we must make our interval wider.)

Let's review what we just did. We wanted to estimate a parameter of the population (the population mean). We were going to use information about a statistic (the sample mean) to make inference about our parameter. To do that, we had to know the distribution of the statistic (distribution of the sample mean, DSM). The distribution gave us information "how close . . . how often." Then we drew our sample and computed the statistic (sample mean, $\bar{x}$).

Now we use the "how close . . . how often" information from the DSM as follows. The "how close" tells us how wide an interval we have to make $\pm$ from the sample mean $\bar{x}$ to capture the population mean. The "how often" tells us how often we will be correct, that is, it tells us the *confidence level* of the estimate.

Using the sample mean to estimate the population mean is a model of all of our estimation-type problems. In each case, we will draw a sample, compute a statistic from the sample, and with our knowledge of the distribution of the statistic, form a confidence interval ("how close . . . how often") about the statistic as an estimator of the parameter. Once we know what the distribution of the sample statistic is, we're in a position to do inferential statistics. The tables at the back of the book give many of the distributions we will need. These distributions have been either worked out theoretically (mathematically) or observed empirically by repeated sampling.

EXERCISE 1

Using our example of tire mileage lifetime, and assuming that $n = 30$, the sample mean $\bar{x} = 38,000$, and sample variance $s^2 = 2,960,689.7$, review the computation of the 90% and 95% confidence intervals; then form 68.3% and 99% confidence intervals for the population mean.

_____

_____

_____

_____

EXERCISE 2

Still using the tire mileage example, and assuming the sample size, sample mean, and sample variance given in Exercise 1, what confidence level could you attach to the interval (37,685.9, 38,314.1)? To the interval (37,057.6, 38,942.4)? To the interval (38,000, 38,000)?

_____

_____

_____

_____

EXERCISE 3

The following data give the fuel economy (miles per gallon) of a sample of 40 automobiles taken from the assembly line and tested.

20	19	20	26	19	22	23	18	18	16
21	19	19	18	21	21	20	20	24	22
19	21	20	20	18	18	19	19	20	20
21	24	23	19	19	20	19	20	19	20

a. Compute the sample variance s^2 and the mean $\bar{x}$.

$s^2 = $ _____ ; $\bar{x} = $ _____ .

b. Using s^2 as an estimate of the population variance, calculate the variance of the DSM. [Use formula (2), page 98. Remember, $n = 40$.]

_____

c. Calculate

$$\sigma_{DSM} \cong \sqrt{\frac{s^2}{n}},$$

the standard deviation of the DSM, from part b.

_____

d. Form 90%, 95%, and 99% confidence intervals for the mean of the population from which this sample was drawn.

_____

_____

_____

"The higher the confidence level, the _____
the interval."

e. Assuming that the sample variance is a good estimate of the population variance, what n would you need if you wanted to form a 99% confidence interval that was only 0.2 wide? [You must choose the sample size that estimates the population mean to within ± 0.1 (that is, 0.2 wide) at the 99% confidence level. Since the only thing under your control is the sample size, look back at formula (2) to see how adjusting the sample size affects the standard deviation of the DSM. How many sigmas of the DSM is a 99% confidence interval? That many sigmas is supposed to be 0.1. So what must sigma be? How large must n be? Remember, n is the sample size. If n is not a whole number, you must go to the next largest integer.]

_____

_____

_____

_____

3-2 ESTIMATION OF PROPORTIONS

Proportions is probably the single most important special case of means that we deal with in a statistical world: the proportion of customers who are dissatisfied with my service; the proportion of defective parts produced in an assembly line; the proportion of cures by a new wonder drug; the proportion of students entering college who finish college in four years.

A **proportion** is a percentage, or fraction, of the whole, and as such is a number between 0 and 1. In the examples mentioned above, and many more that you can think of yourself, the proportion is of considerable interest. Just as we used the _sample mean_ to estimate the _population mean_, we are going to use the _sample proportion_ to estimate the _population proportion p_.

What we need to do first is find a way to attach numbers to our population so that the mean of the numbers turns out to be the proportion of whatever it is we want—such as the proportion of dissatisfied customers or the proportion of defectives. One way to do this is to attach to each individual or object in the population either a 1 or a 0, depending on whether the individual is dissatisfied or not, or whether the part is defective or not. That is, to each dissatisfied customer in the population we attach the number 1; to each satisfied customer we attach a 0. Then if I have 125 dissatisfied customers out of 1000 total customers, I have a collection of 125 1s and 875 0s. The mean of this collection is

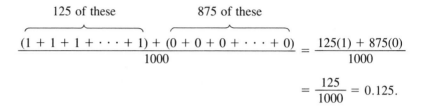

$$= \frac{125}{1000} = 0.125.$$

Thus, the mean or proportion is 0.125, reflecting the fact that 12.5% of the customers are dissatisfied. Of course, we could attach the 0s to the dissatisfied customers and the 1s to the satisfied customers, and then the mean of the collection would be 0.875, reflecting the fact that 87.5% of the population is satisfied. So we want to attach 1s to the kind of customer we wish to measure—or the defective parts, or the cured patients, or the graduating students—and attach 0s to the other kind. Then the mean of the resulting collection of numbers will be the proportion of numbers in the population with the attribute we wish to measure.

A population that consists entirely of 0s and 1s, which is what we have here, is called a **binomial population**. What we have just seen is that the *mean of a binomial population is the proportion of 1s in the population*, which we denote p. Since the proportion of 1s in the population is the mean of the population, we can estimate the proportion just as we estimate any mean: (1) we draw a random sample of some size; (2) we compute the mean of the sample; and (3) we form a confidence interval about the sample mean. Now, just as the population mean is the proportion of 1s in the population, so the sample mean is the proportion of 1s in the sample. So we draw a sample of size n and count the 1s in the sample; let's say there are m 1s in the sample. Then the sample mean = the sample proportion = m/n (denoted $\bar{p}$). The proportion of 0s is denoted $\bar{q}$ ($= 1 - \bar{p}$).

In the general case, we know that the DSM (distribution of sample means) is approximately normal, provided n is 30 or more. In the special case of a binomial population, the DSM is called a *binomial distribution* and tends to be more nearly normal if p, the true population proportion of 1s, is not too near 0 or too near 1. We have a special rule: *we use the normal approximation to the (binomial) DSM, provided the sample we draw has at least 5 0s and at least 5 1s.* (If this isn't the case, we have to use special charts to form confidence intervals for proportions, and these are given in Tables 6a and 6b at the back of the book. I'll tell you about these shortly.)

Let's assume our sample contains at least 5 0s and 5 1s: the mean of the DSM is p (unknown, of course), and it's a mathematical fact that the variance is $\dfrac{pq}{n}$, where q is the proportion of 0s $= 1 - p$ and n is the sample size. Since we don't know p or q, we use $\bar{p}$ and $\bar{q}$ in their place in the calculation of variance. This approximation doesn't significantly affect the calculation. Then σ of the DSM is the square root of the variance, which is approximately

$$\sqrt{\frac{\bar{p}\bar{q}}{n}} = \sqrt{\frac{\bar{p}(1-\bar{p})}{n}} \,.$$

Now we use this σ to calculate a confidence interval for the proportion, just as we did for the mean. A 95% confidence interval for p is

$$\bar{p} \pm 1.96\sigma_{\text{DSM}} = \bar{p} \pm 1.96\sqrt{\frac{\bar{p}\bar{q}}{n}} = \bar{p} \pm 1.96\sqrt{\frac{\bar{p}(1-\bar{p})}{n}} \,.$$

A NORMAL APPROXIMATION EXAMPLE

Suppose I regard my 1000 customers (page 106) as a sample from a potential population (of, hopefully, thousands or millions). Now suppose I want to estimate the true proportion of dissatisfied customers I will have as I continue my business. Here is my computation:

$$n = 1000; \quad \bar{p} = \frac{125}{1000} = 0.125; \quad \bar{q} = \frac{875}{1000} = 0.875;$$

$$\sigma_{\text{DSM}} \cong \sqrt{\frac{\bar{p}\bar{q}}{n}} = \sqrt{\frac{(0.125)(0.875)}{1000}}$$

$$= 0.0104583 \cong 0.0105.$$

Thus, a 95% confidence interval for the proportion p is

$$(\bar{p} \pm 1.96\sigma) = [0.125 \pm 1.96(0.0105)]$$

$$= (0.125 - 0.021,\ 0.125 + 0.021)$$

$$= (0.104,\ 0.146).$$

And, similarly, a 90% confidence interval for the proportion p is

$$(\bar{p} \pm 1.645\sigma) = [0.125 \pm 1.645(0.0105)]$$

$$= (0.125 - 0.017,\ 0.125 + 0.017)$$

$$= (0.107,\ 0.142).$$

We obtained the number 1.645 by referring to the entry 90% in Table 3-1; and we obtained the number 1.96 by referring to the entry 95%. Other values are obtained in a similar manner. Notice that the 90% confidence interval is narrower than the 95% confidence interval, reflecting the fact that when we decrease the likelihood that a statement about the location of the mean is correct, we get a more precise estimate (narrower interval).

USING THE BINOMIAL CHARTS

If our sample doesn't contain at least 5 0s and 5 1s, we look in Table 6a (at the back of the book) for 95% confidence intervals (and in Table 6b for 99% confidence intervals). To see how to use these charts, let's suppose we have 4 1s and 46 0s in our sample. Then

$$\bar{p} = \frac{m}{n} = \frac{4}{50} = 0.08.$$

Now look at Figure 3-7, where I have reproduced the part of the chart in Table 6a corresponding to a sample size of 50. The horizontal axis is $\bar{p}$, the statistic, and the vertical axis is p, the parameter. We observe that $\bar{p} = 0.08$. Then the 95% confidence interval for p is the vertical line segment between the curves marked $n = 50$, whose endpoints are read off at the left: 0.02 and 0.23. Table 6b is similar in layout

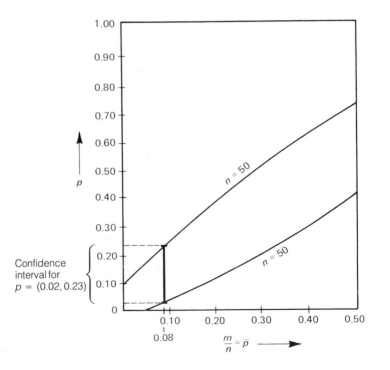

Figure 3-7 95% confidence limits, $n = 50$.

and gives 99% confidence limits. In the actual Tables 6a and 6b at the back of the book, there is a pair of curves just like $n = 50$ for each different sample size.

Notice that the values along the bottom of the chart run only from 0.00 to 0.50. Actually, the values of $\bar{p}$ between 0.50 and 1.00 are placed along the *top* of the same chart and read off from the *right* rather than the left. For example, if we had 4 0s and 46 1s instead, $\bar{p}$ would then be 0.92, and we would compute our 95% confidence interval from the same chart but read from the top and right instead. Look at Figure 3-8, where we do this.

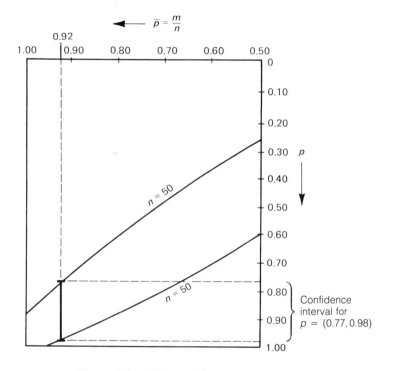

Figure 3-8 95% confidence limits, $n = 50$.

EXERCISE 4

For each of the following sets of data, check whether or not we may use the normal approximation to the binomial:

a. $n = 20$;   8 cured;   12 not cured.

b. $n = 100$;   50 Republicans;   50 Democrats.

c. $n = 58$;   10 satisfied;   48 dissatisfied.

d. $n = 10,000$;   100 defective;   9900 working.

e. $n = 24$;   3 got probation;   21 got jail sentences.

EXERCISE 5   For each of the sets of data in Exercise 4, make up populations from which these samples were drawn, and say what thing is being estimated.

a. _____

_____

b. _____

_____

c. _____

_____

d. _____

_____

e. _____

_____

EXERCISE 6   Form 90%, 95%, and 99% confidence intervals for the population proportions you identified in Exercise 5, parts a–d. Use Tables 6a and 6b to compute 95% and 99% confidence intervals for part e.

a. _____

_____

_____

b. _____

_____

_____

c. _____

_____

_____

d. _____

_____

_____

e. _____

_____

EXERCISE 7   You interview 1453 people by telephone, asking whether or not they will support the new bond issue. 765 say yes, 688 say no. If the election were held today, would the bond issue pass? (Form a 95% confidence interval for the true proportion that support the bond issue, and see if *all* the numbers in the interval are bigger than 50%. If they are, then at the 95% confidence level, the true proportion in favor is greater than 50%, and the bond issue will pass.)

EXERCISE 8   When the air traffic controllers went on strike in 1981, Dan Rather reported on the evening news that 1500 people were polled about whether or not President Reagan was handling the strike properly. 64% of them said that he was. (And 36% said he wasn't.) Rather reported that these figures were accurate to within 3 percentage points, by which he meant that the true figures were 64% ± 3%. He didn't say what his confidence level was (nor did he say that his statement was of confidence interval type), merely that his figures were "subject to a margin of error of 3 percentage points." What must the confidence level of his statement really have been if the confidence interval was 64% ± 3%, with a sample size of 1500?

EXERCISE 9   You wish to conduct a survey to determine the proportion of people who prefer Pepsi to Coke. You want to choose a sample size large enough that you can be 99% confident that your estimate $\bar{p}$ of the proportion lies within ±0.05 = ±5% of the true proportion. What sample size should you choose? (*Hint*: Assume $p = q = 0.50$; this is the "worst case," as it makes for the largest variance and hence the largest width intervals. No matter what the true proportion turns out to be, your interval will be wide enough to make your computation correct.)

QUALITY CONTROL (OPTIONAL)

Here is an example in converting from counts to proportions, and back again. Suppose we are nail manufacturers and wish to put enough nails in each box so that each customer will find at least 1000 good nails in the box, perhaps along with a few defectives. Suppose we know that our production process is such that we produce

98% good nails and 2% defectives. How many nails should we put in each box to ensure that the customer will find 1000 good ones? If we put only 1000 nails in the box, the customer will probably find about 20 defectives (2% of 1000), leaving only 980 good nails.

If you think about it for a moment, you can see that we can *never* be 100% certain that there are 1000 good nails in the box, no matter *how many* nails we put in. Even if we put a million nails in the box, there will be some small probability that all of them are defective. But we can be 95% sure, or 99% sure. We can decide whatever level of certainty we wish and use our statistical machinery to determine the number of nails to put into each box to obtain that level of certainty that there will be at least 1000 good nails in the box.

For example, suppose we wish to be 95% sure that each box has at least 1000 good nails. Then we wish to find an *n* such that when we put *n* nails into the box, the chances of finding fewer than 1000 good nails is less than 5%. As before, we use the binomial distribution and the fact that the binomial distribution is well approximated by the normal. Note that we are working with *counts* of successes and failures rather than with *proportions* of successes and failures. Counts are integers (whole numbers), while proportions are fractions between 0 and 1. But this is only an arithmetic difference. We simply convert to proportions to use our statistical machinery. (We could also work directly with counts, but we would have to introduce additional distributions for that purpose.) Then we convert back to counts. Here is how it works.

We said that we would put *n* nails into the box. The true proportion of good nails, *p*, is 98%, or 0.98. So the variance of the sample proportion, $\bar{p}$, is

$$\sigma^2_{\mathrm{DSM}} = \frac{pq}{n} = \frac{(0.98)(0.02)}{n} = \frac{0.0196}{n} \; .$$

Thus

$$\sigma = \sqrt{\frac{0.0196}{n}} \; .$$

(When we talk about our "sample," we are talking about the one box with *n* nails in it. It is, of course, a sample of size *n*.) Now, the distribution of the sample proportion (the distribution of $\bar{p}$) is a DSM that looks like Figure 3-9.

We are looking for how far we have to go out from the mean in a normal distribution so that we have 5% in *one* tail of the distribution, the 5% of the time that the box will contain fewer than 1000 good nails. We have a one-tailed normal table (see Table 1 at the back of the book), but rather than talk about it at this time, we can use our usual normal table, Table 2, by observing that the distance we have to go to find 5% in the *left tail* is the same as the distance we have to go to find 10% in both tails—5% in the left and 5% in the right. We look in our table for how many sigmas we have to go to find 90%, and we see that that is 1.645σ.

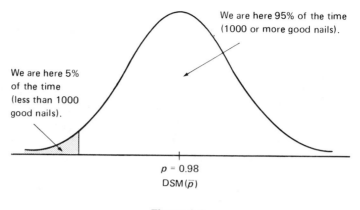

We are here 95% of the time (1000 or more good nails).

We are here 5% of the time (less than 1000 good nails).

$p = 0.98$
$\mathrm{DSM}(\bar{p})$

Figure 3-9

We know that n is approximately 1020 (because 2% defectives out of 1000 is approximately 20), but we are trying to find n more exactly. So

$$\sigma = \sqrt{\frac{0.0196}{n}} \cong \sqrt{\frac{0.0196}{1020}} \cong 0.0044.$$

Then

$$1.645\sigma = 1.645(0.0044) \cong 0.007,$$

and our distribution looks like Figure 3-10. What this means is that for sample size approximately 1020, we will produce less than 97.3% good nails about 5% of the time. So we set the number of nails in the box to be $n = 1000/0.973 \cong 1027.75$.

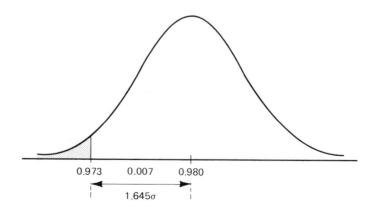

0.973   0.007   0.980

1.645σ

Figure 3-10

Then rounding up to the next whole number, $n = 1028$. A box with 1028 nails will contain fewer than 1000 good nails 5% of the time. (Another way of saying this is: of all the boxes containing 1028 nails, 5% of them will have fewer than 1000 good nails in them.)

If it bothers you that we took n to be 1020 in the calculation of σ, when n turned out to be 1028, we can repeat the calculation of σ with $n = 1028$ and get the same answer to three decimal places; that is, the difference in the value of n makes no significant difference in the calculation of the confidence interval. (Repeat the calculation of σ with $n = 1028$ and write your answer here: $\sigma = $ _____ . (¿Did you get 0.0043665?)

As I said earlier, we can make the complete calculation with count data rather than with proportions. It simply involves knowing what the variance of the distribution is for counts rather than for proportions. In this case, variance $= \sigma^2 = npq$ instead of pq/n. I'll go through the calculation for you, although you don't have to use this method—the method of proportions is sufficient.

$$p = 0.98; \quad q = 0.02; \quad n \cong 1020;$$

$$\text{Variance} = \sigma^2 = npq = (1020)(0.98)(0.02)$$

$$\cong 19.992;$$

$$\sigma \cong 4.47; \quad 1.645\sigma \cong 7.355.$$

If we put n nails into a box, we expect, on average, to have $0.98n$ good ones, with a distribution that looks like Figure 3-11. If we wish there to be at least 1000

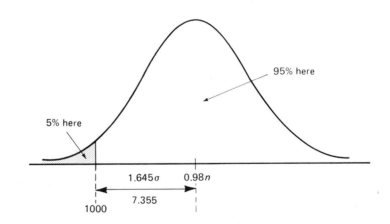

Figure 3-11

good ones, then

$$1000 = 0.98n - 1.645\sigma$$
$$= 0.98n - 7.355$$
$$1007.355 = 0.98n$$
$$n = \frac{1007.355}{0.98} = 1027.91 \cong 1028.$$

EXERCISE 10 You have a manufacturing process and employ quality-control supervisors to ensure that your finished product has fewer than 3% defectives. Yet in one week, when you shipped 2000 of your product, you got complaints about 100 defectives. Is this reasonable? What is the likelihood, if your process has 3% or fewer defectives, that you will observe 100 defectives or more out of a batch of 2000?

_____

_____

_____

3-3 DIFFERENCES BETWEEN MEANS (INDEPENDENT SAMPLES)

Are brand A tires better than brand B? How would you measure? Well, most people would take the tire that got the better mileage, all other things being equal. So if brand A tires average 38,000 miles per tire, and brand B tires average 40,000 miles per tire, then most people would prefer brand B to brand A. The difference between their means is $40,000 - 38,000 = 2000$ miles per tire, with the plus side going to brand B. Clearly, the *difference between means* is of considerable interest to any manufacturer whose product can be measured by a single number.

To place things in a framework that we already understand, let us consider two populations, population A and population B, with means μ_A and μ_B, respectively. What we want to know is the difference between μ_A and μ_B, or $\mu_A - \mu_B$. This is a parameter of the two populations, and as you might expect, the statistic we will use to estimate $\mu_A - \mu_B$ is the difference between the means of two samples, $\bar{x}_A - \bar{x}_B$. Thinking about how we used one sample mean to estimate one population mean, we'll use the *difference* between sample means to estimate the *difference* between population means in the same way. We must assume that the two samples are drawn independently of one another.

First we need to know the distribution of the statistic: the difference between the sample means. (We'll call the distribution of differences between sample means

DDSM for short.) All we need to know about the DDSM is the following:

1. The mean of the DDSM is the difference between the population means, the parameter we are trying to estimate.

2. The DDSM is approximately normal if the sizes of the samples from population A and population B are both of size 30 or more.

3. The variance of the DDSM is the *sum* of the variances of the DSM for $\bar{x}_A$ and the DSM for $\bar{x}_B$.

Now we're all set to do statistical inference for the tire manufacturers. Suppose we draw a sample of size 30 from population A and get a sample mean (tire lifetime) of 38,000. And suppose we draw a sample of size 40 from population B and get a sample mean of 40,000. And suppose also that the sample variance for sample A is 1,000,000 and that the sample variance for sample B is 2,000,000. To form a 95% confidence interval for the difference between population means, we go out $\pm 1.96\sigma_{DDSM}$ from the difference between sample means:

$$38,000 - 40,000 = -2000.$$

(The minus sign indicates that the mean tire mileage from sample A is less than the mean tire mileage from sample B. We could compute sample mean B − sample mean A and get +2000 miles. So the minus sign is arbitrary.) To determine the interval, we need only determine $1.96\sigma_{DDSM}$, which means that (once again) we need σ of the distribution. According to item 3 in the list, the variance of the DDSM is the sum of the variances of the DSMs for $\bar{x}_A$ and $\bar{x}_B$. The variance of the DSM for $\bar{x}_A$, $\sigma^2_{\bar{x}_A}$, is the variance of population A ÷ 30, the sample size. Our estimate for the variance of population A is the variance of sample A, 1,000,000. Therefore, $\sigma^2_{\bar{x}_A} \cong$ 1,000,000/30. Similarly, $\sigma^2_{\bar{x}_B} \cong$ 2,000,000/40. Thus, our value for the variance of the DDSM is the sum of the variances of the two DSMs, and so σ of the DDSM is

$$\sigma_{DDSM} \cong \sqrt{\frac{1,000,000}{30} + \frac{2,000,000}{40}} \tag{3}$$

$$= \sqrt{33,333.3 + 50,000}$$

$$= \sqrt{83,333.3}$$

$$= 288.68.$$

So $1.96\sigma = (1.96)(288.67) = 565.8$, and our 95% confidence interval for the difference between population means is

$$(-2000 \pm 1.96\sigma) = (-2000 \pm 565.8) = (-2565.8, -1434.2).$$

Note that according to this, the chances are less than 5% that the true difference between means is 0 or greater (since 0 is outside the 95% confidence interval). Thus, the manufacturer of tire *B* would be justified in saying that these data support his claim at the 95% confidence level that his tires get better mileage than brand *A* tires. In Chapter 4 I'll show you how we **test** such a claim directly without confidence intervals.

When we made our calculation of the standard deviation of the DDSM, we used the fact that the variance of the DDSM was the sum of the variances of the two DSMs and that the variance of each DSM was the variance of the respective population divided by the respective sample size. So recalling that we used the variance of each sample as the value of the variance of each population, we computed the following:

$$\sigma_{\text{DDSM}} \cong \sqrt{\frac{\text{Variance (sample } A)}{\text{Size (sample } A)} + \frac{\text{Variance (sample } B)}{\text{Size (sample } B)}}$$

$$= \sqrt{\frac{s_A^2}{n} + \frac{s_B^2}{m}}, \tag{4}$$

which came from

$$\text{Variance(DDSM)} = \text{Variance(DSM}_{\bar{x}_A}) + \text{Variance(DSM}_{\bar{x}_B})$$

and

$$\text{Variance(DSM}_{\bar{x}_A}) = \frac{\text{var(pop } A)}{n} \cong \frac{\text{var(sample } A)}{n}$$

or

$$\frac{\sigma_{\text{pop } A}^2}{n} \cong \frac{s_A^2}{n}$$

and

$$\text{Variance(DSM}_{\bar{x}_B}) = \frac{\text{var(pop } B)}{m} \cong \frac{\text{var(sample } B)}{m}$$

or

$$\frac{\sigma_{\text{pop } B}^2}{m} \cong \frac{s_B^2}{m},$$

where n = sample size from population A and m = sample size from population B. Of course, the variance s_A^2 of sample A is only approximately the variance σ_A^2 of population A, and so on. But as before, this isn't a problem.

It's okay if you don't care where equation (3) comes from. All you have to know is how to use it. In our example above, we had a sample of size 30 from population A, with a sample variance of 1,000,000; we had a sample of size 40 from population B, with a sample variance of 2,000,000. We just computed (3) according to equation (4). Compare (3) with (4) to see this. Every case is just as easy as this. Of course, once we have σ_{DDSM}, we form our confidence interval as we always do: the value of the statistic $\bar{x}_A - \bar{x}_B$, $\pm$ so many sigmas of the DDSM.

EXERCISE 11

Use the two collections of numbers in Examples 1 and 2 from Chapter 1, page 9, the ages of kings and queens at coronation, and the ages of U.S. presidents at inauguration, as samples from some larger populations. Compute a 95% confidence interval for the difference between population means of these two populations. Ignore the fact that Example 1 has an n of 20. Use formula (4) anyway. Remember to calculate *sample* variances.

3-4 DIFFERENCES BETWEEN PROPORTIONS

The more we do this, the easier it gets! This next topic should really be easy for you, now that you've had so much practice. We're going to estimate the difference between proportions in two binomial populations. The setup could be the following. Suppose we have two headache remedies, drug A and drug B. Is A better than B? Presumably, we are asking if A cures (proportionately) more headaches than B does. Or, saying the same thing another way: Is A's *cure rate* better than B's?

We can think of this problem in terms of binomial populations. Consider all the people in the world who ever have had, or will have, a headache. Then treat them with drug A. If they get better (cured), attach a 1 to them. Otherwise, attach a 0. This is binomial population A. Now (hypothetically) repeat the process with the *same people*, but with drug B. Then attach 1s to those people cured by B, and 0s otherwise. This is population B. (Don't worry about the fact that much of this population is hypothetical, in that some of the people who contribute 0s or 1s to population A or B haven't been born yet or don't have a headache yet. This is only a theoretical problem, and practical issues won't stand in the way of our doing solid statistical inference.)

Now we have two binomial populations, population A and population B. Again, to say that drug A is better than drug B, presumably we mean that A's cure rate is better (greater) than B's cure rate. And that's just the proportion of 1s in population A versus the proportion of 1s in population B. To say that A is better than B, we mean that the difference between means (proportions, in this case), $p_A - p_B$, is positive (meaning p_A is greater than p_B).

To determine a 95% confidence interval for $p_A - p_B$, the difference between population proportions, we choose a sample of size n from population A and a sample of size m from population B. (Usually we choose the same size samples, but we can imagine circumstances where we have different sample sizes.) Now we count the proportion of 1s in each sample, which is in each case a fraction, say $\bar{p}_A$ and $\bar{p}_B$. Then (as usual) we take

$$(\bar{p}_A - \bar{p}_B) \pm 1.96\sigma_{\text{DDSP}},$$

where σ_{DDSP} is the standard deviation of the distribution of differences between sample proportions, DDSP (just like the distribution of differences between sample means, DDSM). To complete the calculation, we need to know the following:

1. The DDSP is approximately normal, provided the total number of 0s observed is at least 5 and the total number of 1s is at least 5.

2. The variance of the DDSP is the sum of the variances of the DSMs for sample proportions in each of the two populations.

Point 1 means that when we *pool* the observations drawn from samples A and B, there are a total of at least 5 0s and 5 1s. Point 2 is just like point 3 in Section 3-3 (page 116); that is, the variance of the sum of two (independent random variables) statistics is just the sum of the variances of each of them. Recall from page 107 that

$$\sigma^2_{\text{DSM}} = \frac{p(1 - p)}{n} \cong \frac{\bar{p}(1 - \bar{p})}{n},$$

where $\bar{p}$ is the proportion of 1s in the sample and n is the sample size. If this is the (approximate) variance for sample A from population A, and the variance for sample

B from population B is similarly approximated, then the variance of the DDSP is their sum, and σ_{DDSP} is given by

$$\sigma_{\text{DDSP}} \cong \sqrt{\frac{\bar{p}_A(1 - \bar{p}_A)}{n} + \frac{\bar{p}_B(1 - \bar{p}_B)}{m}}, \tag{5}$$

where $\bar{p}_A$ is the proportion of 1s in sample A, $\bar{p}_B$ is the proportion of 1s in sample B, n is the size of sample A, and m is the size of sample B. Compare formula (5) with formula (4) and you can see how similar they are.

AN EXAMPLE OF DIFFERENCES BETWEEN PROPORTIONS

Suppose we give drug A to 100 people with headaches, and 50 out of 100 are cured. And suppose we give drug B to 200 people with headaches, and 60 out of 200 are cured. Can we say drug A is better than drug B (because it cured a higher proportion of headaches)? Let's form a 95% confidence interval for the difference between means and see. Since $\bar{p}_A = 50/100 = 0.50$ and $\bar{p}_B = 60/200 = 0.30$,

$$\sigma_{\text{DDSP}} \cong \sqrt{\frac{(0.50)(1 - 0.50)}{100} + \frac{(0.30)(1 - 0.30)}{200}}$$

$$= 0.05958.$$

Then a 95% confidence interval for the true difference between proportions is $(0.50 - 0.30) \pm 1.96\sigma$ or

$$0.20 \pm 1.96(0.06) = (0.20 \pm 0.1168)$$

$$= (0.083, \ 0.317).$$

Note that the 95% confidence interval for the true difference does not include either 0 or any negative values, strengthening the claim that A is better than B (by between 8.3% and 31.7%). And notice also that the 99% confidence interval,

$$0.20 \pm 2.576(0.06) = (0.047, \ 0.353)$$

also does not include 0 or any negative values. In Chapter 4 we'll present a method designed specifically to test whether or not two means, or two proportions, are the same.

EXERCISE 12

Suppose that 40 out of 100 individuals treated with drug A are cured, and 60 out of 200 treated with drug B are cured. Thus, drug A has a 40% cure rate, while drug B

has a 30% cure rate. Can we say drug A is better than drug B at the 90% confidence level? (See the above example for the method to use to solve this problem.)

_____

_____

_____

PROBLEM SET A

1. Regarding the following as a sample of 50 corporation presidents' ages, compute
   a. the standard deviation of the distribution of sample means, as we did in Exercise 3
   b. a 90% confidence interval for the mean
   c. a 95% confidence interval for the mean
   d. a 99% confidence interval for the mean

58	59	63	61	62	47	56	62	63	51
69	58	67	68	59	67	60	60	54	63
54	67	66	57	65	57	55	58	65	51
59	70	63	59	74	71	58	51	43	44
61	59	61	70	61	63	54	60	60	71

2. Regarding the following as a sample of 26 daily expenses, compute
   a. the standard deviation of the distribution of sample means, as we did in Exercise 3
   b. a 90% confidence interval for the mean
   c. a 95% confidence interval for the mean
   d. a 99% confidence interval for the mean

44.98	66.56	49.28	68.03	54.95	58.49	54.68	51.37
39.06	50.05	47.57	61.20	72.60	77.11	41.24	47.97
71.35	42.91	65.03	57.54	43.54	69.17	45.39	38.89
59.97	36.43						

3. Regarding the following as a sample of 45 male models' heights, in inches, compute
   a. the standard deviation of the distribution of sample means, as we did in Exercise 3
   b. a 90% confidence interval for the mean
   c. a 95% confidence interval for the mean
   d. a 99% confidence interval for the mean

71	72	70	71	72	72	70	70	71	70
73	72	70	72	70	74	72	72	73	71
71	72	71	73	72	71	72	71	74	71
71	71	74	73	72	71	72	70	71	71
73	70	70	72	72					

4. Regarding the following as a sample of 15 students' attention spans, in minutes, compute
   a. the standard deviation of the distribution of sample means, as we did in Exercise 3
   b. a 90% confidence interval for the mean
   c. a 95% confidence interval for the mean
   d. a 99% confidence interval for the mean

10.9	10.1	9.2	13.6	9.9	9.7	13.8	13.0	5.0	6.6
8.6	14.0	8.5	9.9	10.1					

5. Regarding the following as a sample of 30 automobile gasoline mileages, compute
   a. the standard deviation of the distribution of sample means, as we did in Exercise 3
   b. a 90% confidence interval for the mean
   c. a 95% confidence interval for the mean
   d. a 99% confidence interval for the mean

21.9	38.6	34.7	31.2	17.2	18.6	28.4	30.3	35.7
21.9	23.1	38.8	38.0	37.3	16.2	22.7	28.3	22.4
24.0	15.6	31.5	22.7	24.4	16.9	26.6	21.8	30.6
19.6	35.8	23.0						

6. Regarding the following as a sample of 35 television repair bills, compute
   a. the standard deviation of the distribution of sample means, as we did in Exercise 3
   b. a 90% confidence interval for the mean
   c. a 95% confidence interval for the mean
   d. a 99% confidence interval for the mean

60.33	55.82	55.99	165.58	96.48	154.74	137.06
108.82	67.42	79.94	87.44	159.72	88.50	175.32
165.41	153.53	104.00	178.08	49.35	65.01	119.59
121.20	130.35	130.98	154.85	155.94	124.60	52.77
75.76	94.34	126.01	158.76	107.22	111.72	98.59

7. Regarding the following as a sample of 15 rats' weight gains, in grams, compute
   a. the standard deviation of the distribution of sample means, as we did in Exercise 3
   b. a 90% confidence interval for the mean
   c. a 95% confidence interval for the mean
   d. a 99% confidence interval for the mean

8.8	28.3	9.6	9.6	42.0	43.3	5.3	19.6	34.2	13.2
4.2	45.9	5.9	12.6	10.7					

8. Regarding the following as a sample of 32 executive salaries, compute

    a. the standard deviation of the distribution of sample means, as we did in Exercise 3
    b. a 90% confidence interval for the mean
    c. a 95% confidence interval for the mean
    d. a 99% confidence interval for the mean

128,000	145,000	78,000	136,000	60,000	72,000	91,000
217,000	46,000	191,000	85,000	127,000	139,000	134,000
182,000	104,000	173,000	177,000	197,000	213,000	149,000
112,000	133,000	145,000	101,000	155,000	167,000	143,000
213,000	88,000	149,000	173,000			

9. Regarding the following as a sample of 60 pocketsful of change, in cents, compute

    a. the standard deviation of the distribution of sample means, as we did in Exercise 3
    b. a 90% confidence interval for the mean
    c. a 95% confidence interval for the mean
    d. a 99% confidence interval for the mean

7	18	0	89	4	24	6	21	21	15
19	45	55	71	32	29	54	62	33	38
11	79	69	51	86	76	60	63	28	72
0	32	57	18	85	27	63	48	58	37
86	25	15	74	5	21	62	26	68	1
87	81	84	71	1	78	83	30	8	69

10. Regarding the following as a sample of 30 flying times (in minutes) to cross the United States, compute

    a. the standard deviation of the distribution of sample means, as we did in Exercise 3
    b. a 90% confidence interval for the mean
    c. a 95% confidence interval for the mean
    d. a 99% confidence interval for the mean

334	317	290	289	285	264	294	280	324	301
306	316	286	328	358	276	358	308	325	354
289	281	253	268	328	270	268	318	289	332

11. Regarding the data at the top of page 124 as a sample of 33 measurements of galvanic skin response, in volts, compute

    a. the standard deviation of the distribution of sample means, as we did in Exercise 3

b. a 90% confidence interval for the mean
c. a 95% confidence interval for the mean
d. a 99% confidence interval for the mean

0.0213	0.0182	0.0201	0.0188	0.0204	0.0181	0.0190
0.0198	0.0214	0.0204	0.0208	0.0206	0.0210	0.0206
0.0200	0.0215	0.0206	0.0199	0.0206	0.0210	0.0190
0.0192	0.0202	0.0218	0.0181	0.0217	0.0180	0.0180
0.0182	0.0201	0.0197	0.0190	0.0214		

PROBLEM SET B

1. A sample of 1713 individuals is tested. 1496 of the individuals respond positively. Estimate the true proportion of the population that would respond positively

   a. at the 90% confidence level
   b. at the 95% confidence level
   c. at the 99% confidence level

2. A sample of 127 individuals is polled. 101 of the individuals are in favor. Estimate the true proportion of the population that is in favor

   a. at the 90% confidence level
   b. at the 95% confidence level
   c. at the 99% confidence level

3. A sample of 539 individuals is tested. 188 of the individuals respond positively. Estimate the true proportion of the population that would respond positively

   a. at the 90% confidence level
   b. at the 95% confidence level
   c. at the 99% confidence level

4. A sample of 1476 individuals is polled. 1050 of the individuals are in favor. Estimate the true proportion of the population that is in favor

   a. at the 90% confidence level
   b. at the 95% confidence level
   c. at the 99% confidence level

5. A sample of 1954 individuals is tested. 292 of the individuals respond positively. Estimate the true proportion of the population that would respond positively

   a. at the 90% confidence level
   b. at the 95% confidence level
   c. at the 99% confidence level

6. A sample of 1648 individuals is polled. 417 of the individuals are in favor. Estimate the true proportion of the population that is in favor

   a. at the 90% confidence level
   b. at the 95% confidence level
   c. at the 99% confidence level

7. A sample of 411 individuals is tested. 63 of the individuals respond positively. Estimate the true proportion of the population that would respond positively
   a. at the 90% confidence level
   b. at the 95% confidence level
   c. at the 99% confidence level

8. A sample of 1962 individuals is polled. 993 of the individuals are in favor. Estimate the true proportion of the population that is in favor
   a. at the 90% confidence level
   b. at the 95% confidence level
   c. at the 99% confidence level

9. A sample of 1044 individuals is tested. 239 of the individuals respond positively. Estimate the true proportion of the population that would respond positively
   a. at the 90% confidence level
   b. at the 95% confidence level
   c. at the 99% confidence level

10. A sample of 1875 individuals is polled. 657 of the individuals are in favor. Estimate the true proportion of the population that is in favor
    a. at the 90% confidence level
    b. at the 95% confidence level
    c. at the 99% confidence level

11. A sample of 200 individuals is tested. 82 of the individuals respond positively. Estimate the true proportion of the population that would respond positively
    a. at the 90% confidence level
    b. at the 95% confidence level
    c. at the 99% confidence level

12. A sample of 1757 individuals is polled. 512 of the individuals are in favor. Estimate the true proportion of the population that is in favor
    a. at the 90% confidence level
    b. at the 95% confidence level
    c. at the 99% confidence level

13. A sample of 469 individuals is tested. 181 of the individuals respond positively. Estimate the true proportion of the population that would respond positively
    a. at the 90% confidence level
    b. at the 95% confidence level
    c. at the 99% confidence level

14. A sample of 176 individuals is polled. 94 of the individuals are in favor. Estimate the true proportion of the population that is in favor
    a. at the 90% confidence level
    b. at the 95% confidence level
    c. at the 99% confidence level

15. A sample of 218 individuals is tested. 34 of the individuals respond positively. Estimate the true proportion of the population that would respond positively
    a. at the 90% confidence level
    b. at the 95% confidence level
    c. at the 99% confidence level

16. A sample of 1680 individuals is polled. 565 of the individuals are in favor. Estimate the true proportion of the population that is in favor
    a. at the 90% confidence level
    b. at the 95% confidence level
    c. at the 99% confidence level

17. A sample of 1620 individuals is tested. 50 of the individuals respond positively. Estimate the true proportion of the population that would respond positively
    a. at the 90% confidence level
    b. at the 95% confidence level
    c. at the 99% confidence level

18. A sample of 1076 individuals is polled. 574 of the individuals are in favor. Estimate the true proportion of the population that is in favor
    a. at the 90% confidence level
    b. at the 95% confidence level
    c. at the 99% confidence level

19. A sample of 1186 individuals is tested. 953 of the individuals respond positively. Estimate the true proportion of the population that would respond positively
    a. at the 90% confidence level
    b. at the 95% confidence level
    c. at the 99% confidence level

20. A sample of 211 individuals is polled. 89 of the individuals are in favor. Estimate the true proportion of the population that is in favor
    a. at the 90% confidence level
    b. at the 95% confidence level
    c. at the 99% confidence level

21. A sample of 795 individuals is tested. 173 of the individuals respond positively. Estimate the true proportion of the population that would respond positively
    a. at the 90% confidence level
    b. at the 95% confidence level
    c. at the 99% confidence level

22. A sample of 1255 individuals is polled. 1020 of the individuals are in favor. Estimate the true proportion of the population that is in favor
    a. at the 90% confidence level
    b. at the 95% confidence level
    c. at the 99% confidence level

23. A sample of 1229 individuals is tested. 818 of the individuals respond positively. Estimate the true proportion of the population that would respond positively

   a. at the 90% confidence level
   b. at the 95% confidence level
   c. at the 99% confidence level

24. A sample of 482 individuals is polled. 170 of the individuals are in favor. Estimate the true proportion of the population that is in favor

   a. at the 90% confidence level
   b. at the 95% confidence level
   c. at the 99% confidence level

25. A sample of 198 individuals is tested. 27 of the individuals respond positively. Estimate the true proportion of the population that would respond positively

   a. at the 90% confidence level
   b. at the 95% confidence level
   c. at the 99% confidence level

26. A sample of 550 individuals is polled. 66 of the individuals are in favor. Estimate the true proportion of the population that is in favor

   a. at the 90% confidence level
   b. at the 95% confidence level
   c. at the 99% confidence level

27. A sample of 879 individuals is tested. 390 of the individuals respond positively. Estimate the true proportion of the population that would respond positively

   a. at the 90% confidence level
   b. at the 95% confidence level
   c. at the 99% confidence level

28. A sample of 1284 individuals is polled. 565 of the individuals are in favor. Estimate the true proportion of the population that is in favor

   a. at the 90% confidence level
   b. at the 95% confidence level
   c. at the 99% confidence level

29. A sample of 940 individuals is tested. 596 of the individuals respond positively. Estimate the true proportion of the population that would respond positively

   a. at the 90% confidence level
   b. at the 95% confidence level
   c. at the 99% confidence level

30. A sample of 955 individuals is polled. 924 of the individuals are in favor. Estimate the true proportion of the population that is in favor

   a. at the 90% confidence level
   b. at the 95% confidence level
   c. at the 99% confidence level

PROBLEM SET C   For each of the following problems, compute

a. a 90% confidence interval

b. a 95% confidence interval

c. a 99% confidence interval

for the difference between means of population *A* and population *B*

1. Sample from population *A*—sample size = 30:

501	494	496	500	492	501	494	504	504	506
482	489	498	496	490	492	484	494	496	487
494	492	506	499	502	513	517	507	498	484

Sample from population *B*—sample size = 35:

301	295	329	311	294	294	307	305	299	287
316	301	299	306	306	291	298	285	311	292
316	295	292	318	308	284	305	304	301	286
310	303	325	302	285					

2. Sample from population *A*—sample size = 36:

59.93	36.47	61.26	49.57	55.40	43.35	57.56	46.58	46.27
51.19	41.73	45.95	58.39	67.74	44.42	75.46	53.27	44.23
58.70	73.47	60.68	43.89	43.30	31.60	75.49	36.76	35.09
62.63	63.80	33.76	71.82	48.73	78.33	76.92	63.10	42.79

Sample from population *B*—sample size = 30:

55.95	73.00	56.63	53.19	65.61	77.26	57.69	71.62	52.42
73.48	57.87	57.36	51.03	62.32	73.07	91.62	75.19	78.78
64.71	70.24	60.68	79.08	75.24	70.61	78.36	74.66	58.84
68.87	58.56	53.48						

3. Sample from population *A*—sample size = 50:

65	73	58	55	70	60	70	60	54	56
56	59	59	66	65	43	67	45	56	54
57	53	63	60	58	73	55	53	61	52
68	63	53	67	54	63	59	59	49	55
61	47	57	63	70	58	50	66	56	50

Sample from population *B*—sample size = 40:

150	131	147	133	132	136	133	136	131	143
144	129	139	134	134	143	146	142	137	151
133	138	143	146	145	142	150	147	142	133
140	131	146	150	135	150	135	143	130	148

4. Sample from population *A*—sample size = 45:

73	70	73	71	72	72	72	72	72	71

71	71	72	71	72	71	72	71	70	72
72	70	70	73	70	73	71	71	70	72
71	72	70	73	72	72	73	71	71	72
72	71	71	73	72					

Sample from population B—sample size = 30:

72	70	72	72	71	71	73	72	71	71
72	71	71	72	71	72	71	71	73	72
73	71	71	72	70	71	71	73	70	73

5. Sample from population A—sample size = 35:

15.1	15.8	16.5	20.7	16.3	23.8	20.0	20.1	20.5	16.7
17.5	21.8	15.4	24.3	19.9	21.0	21.1	19.8	18.6	21.5
21.8	23.3	23.9	20.9	20.7	18.4	15.5	17.8	16.9	21.1
16.0	22.8	24.8	21.3	16.6					

Sample from population B—sample size = 35:

15.5	11.2	11.4	10.3	19.6	10.1	13.1	12.9	11.3	19.2
16.0	16.1	18.6	19.8	19.2	13.2	19.3	14.0	10.2	12.7
16.4	18.6	11.0	18.7	15.4	12.8	17.4	14.6	14.5	17.4
10.3	18.5	17.5	16.8	19.5					

6. Sample from population A—sample size = 30:

19.3	36.9	24.2	15.2	38.2	16.8	29.7	18.6	24.3	26.2
30.4	36.7	15.8	20.1	24.8	31.3	29.3	19.1	15.2	28.9
36.5	28.7	24.4	18.5	22.3	39.3	20.3	29.6	24.4	33.0

Sample from population B—sample size = 40:

20.3	17.3	29.0	21.6	28.2	28.4	22.7	15.8	26.4	23.3
25.7	25.0	25.0	29.7	31.3	39.9	16.5	27.5	26.5	19.6
37.8	39.6	19.2	35.9	36.8	29.3	21.0	24.2	22.2	28.7
37.0	39.4	18.0	23.7	36.5	19.2	24.9	19.6	29.6	28.4

7. Sample from population A—sample size = 35:

228.79	233.33	149.77	200.24	119.44	140.00	133.87
215.65	210.14	137.17	174.57	196.86	156.55	187.83
232.95	230.68	108.25	213.24	235.21	188.45	221.23
115.42	165.25	88.54	210.38	159.72	197.09	178.84
222.01	117.23	227.42	99.30	146.31	207.33	88.09

Sample from population B—sample size = 40:

104.36	29.90	100.19	58.59	65.54	28.29	179.27
118.60	153.84	118.42	163.78	80.42	82.33	131.10
131.24	41.52	168.60	163.58	159.94	97.03	97.27
131.99	71.44	152.23	122.48	163.90	173.12	95.06
70.42	70.74	120.59	168.63	141.80	140.39	60.27
103.67	154.37	49.61	51.87	51.10		

CHAPTER 3 QUIZ

1. "The population mean lies between 950 and 1100 90% of the time." This is a _____ statement.

2. To be able to make a statement about the population mean, you need to know the _____ of the _____ .

3. In general, to make a statement about *some* population parameter, you need to know the _____ of the _____ .

4. For sample means of sample size 35, the variance of the DSM is the population variance _____ .

5. It's a fundamental fact that if the sample size is 30 or more, the DSM is approximately _____ .

6. A proportion is a _____ of the whole.

7. When we study proportions, we use the _____ distribution.

8. Cure rate is an example of _____ .

ANSWERS TO EXERCISES

1. 68.3% confidence interval:

$$\bar{x} \pm 1\sigma_{\text{DSM}} = \bar{x} \pm 1.0\left(\frac{s}{\sqrt{n}}\right) = \bar{x} \pm 1.0(314.1)$$

$$= 38,000 \pm 314.1$$

$$= (37,685.9, \ 38,314.1)$$

99% confidence interval:

$$\bar{x} \pm 2.576\sigma_{\text{DSM}} = \bar{x} \pm 2.576(314.1)$$

$$= (37,191.1, \ 38,808.9)$$

2. $(37,685.9, \ 38,314.1) = 38,000 \pm 314.1$

$$= 38,000 \pm 1.0\sigma_{\text{DSM}}$$

So 68%, from Table 2.

$$(37,057.6, \ 38,942.4) = 38,000 \pm 942.4$$

$$= 38,000 \pm 3.0\sigma_{\text{DSM}}$$

So 99.7% from Table 2.

$$(38{,}000,\ 38{,}000) = 38{,}000 \pm 0 = 38{,}000 \pm 0\sigma_{\text{DSM}}$$

So 0% from Table 2.

3. a. s^2 (sample variance, so s_{n-1}^2) = 3.7333333; $\bar{x} = 20.1$

   b. $\dfrac{s^2}{40} = 0.093333333$

   c. $\dfrac{s}{\sqrt{40}} = \sqrt{0.09333333} = 0.305505$

   d. 90%: $\bar{x} \pm 1.645(0.305505) = 20.1 \pm 0.5$
   $$= (19.6,\ 20.6)$$
   95%: $\bar{x} \pm 1.96(0.305505) = 20.1 \pm 0.6$
   $$= (19.5,\ 20.7)$$
   99%: $\bar{x} \pm 2.576(0.305505) = 20.1 \pm 0.8$
   $$= (19.3,\ 20.9)$$
   Wider.

   e. Here's how to solve the problem. A 99% confidence interval is $\bar{x} \pm 2.576\sigma_{\text{DSM}}$. If the interval is to be 0.2 wide, then $2.576\sigma_{\text{DSM}} = 0.1$. But $\sigma_{\text{DSM}} = s/\sqrt{n}$. So set $2.576s/\sqrt{n} = 0.1$ and solve for n. (We computed s as $1.932+$.) $n = 2477.36$, so choose $n = 2478$.

4. The normal approximation applies to all but part e. The rule is that there must be at least 5 in each category.

5. a. Patients with a certain blood disease. They are either cured or not cured if treated with a new drug. Estimate the proportion of cures, or *cure rate*, if all persons with this disease were treated with the drug.
   b. The population of all registered voters in a certain town. Estimate the proportion of Republicans and the proportion of Democrats in the town.
   c. The population of all customers at my store. Estimate the proportion of satisfied customers.
   d. The population of all thumbtacks produced by my plant. Estimate the proportion of defective thumbtacks.
   e. All persons appearing before Judge Smith and found guilty. Estimate the proportion of such persons who get jail sentences.

6. a. $\sigma_{\text{DSM}} = \sqrt{\dfrac{(^8/_{20})(1 - ^8/_{20})}{20}} = 0.1095$

   90%: $8/20 \pm 1.645(0.1095) = (0.22,\ 0.58)$
   95%: $8/20 \pm 1.96(0.1095) = (0.19,\ 0.61)$
   99%: $8/20 \pm 2.576(0.1095) = (0.12,\ 0.68)$

   b. $\sigma_{\text{DSM}} = \sqrt{\dfrac{(^{50}/_{100})(1 - ^{50}/_{100})}{100}} = 0.05$

   90%: $50/100 \pm 1.645(0.05) = (0.42,\ 0.58)$
   95%: $50/100 \pm 1.96(0.05) = (0.40,\ 0.60)$
   99%: $50/100 \pm 2.576(0.05) = (0.37,\ 0.63)$

c. $\sigma_{DSM} = \sqrt{\dfrac{(^{10}/_{58})(1 - {}^{10}/_{58})}{58}} \cong 0.05$

90%: $10/58 \pm 1.645(0.05) = (0.09, 0.25)$
95%: $10/58 \pm 1.96(0.05) = (0.07, 0.27)$
99%: $10/58 \pm 2.576(0.05) = (0.04, 0.30)$

d. $\sigma_{DSM} = \sqrt{\dfrac{(0.01)(1 - 0.01)}{10,000}} \cong 0.001$

90%: $100/10,000 \pm 1.645(0.001) = (0.008, 0.012)$
95%: $100/10,000 \pm 1.96(0.001) = (0.008, 0.012)$ (to three places)
99%: $100/10,000 \pm 2.576(0.001) = (0.007, 0.013)$

e. 95%: $(0.03, 0.32)$
   99%: $(0.02, 0.38)$

7. $\sigma_{DSM} = \sqrt{\dfrac{(^{765}/_{1453})(1 - {}^{765}/_{1453})}{1453}} \cong 0.013; \quad \bar{p} = \dfrac{765}{1453} = 0.5265$

95%: $0.5265 \pm 1.96(0.013) = (0.501, 0.552)$. At the 95% confidence interval the true proportion of voters in favor is greater than 50%, so the bond issue will pass.

8. $\sigma \cong \sqrt{\dfrac{\bar{p}\bar{q}}{n}} = \sqrt{\dfrac{(0.64)(0.36)}{1500}} = 0.0124$

Then

$$0.03 = \left(\dfrac{0.03}{0.0124}\right)(0.0124) = (2.42)(0.0124) = 2.42\sigma.$$

Looking in our Table 2, we see there is no entry that is *exactly* 2.42; the nearest entry is 2.40, which is the 98.4% confidence level. Actually, what was computed was a 99% confidence interval: $64\% \pm 2.576\sigma = 64\% \pm 2.576(0.0124) = 64\% \pm 3.19\%$, then the "margin of error" was rounded to 3%.

9. The 99% confidence interval is $\bar{p} \pm 2.576\sigma_{DSM}$. If we want within ± 0.05, then

$$2.576\sigma_{DSM} \leqslant 0.05 \qquad \text{or} \qquad 2.576\sqrt{\dfrac{p(1 - p)}{n}} \leqslant 0.05,$$

$$\text{or} \qquad 2.576\sqrt{\dfrac{(0.50)(0.50)}{n}} \leqslant 0.05,$$

$$\text{or} \qquad 2.576\dfrac{\sqrt{(0.50)(0.50)}}{0.05} \leqslant \sqrt{n},$$

$$\text{or} \qquad 25.76 \leqslant \sqrt{n},$$

$$\text{or} \qquad 663.58 \leqslant n.$$

So choose $n = 664$. No matter that $\bar{p}$ turns out to be, you are guaranteed that your interval width will be less than ± 0.05.

10. Here is the DSM, after converting to proportions:

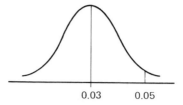

0.03   0.05

The assumption is that p, the true proportion of defectives, is 0.03 *or less*. We set the mean of the DSM as 0.03, because if p were less, it would only make the $\bar{p} = 0.05$ that we observe *even further* from the mean.

$$\sigma_{\text{DSM}} = \sqrt{\frac{(0.03)(0.97)}{2000}} \cong 0.0038$$

Thus 0.05, converted to a z-score, is

$$z = \frac{0.05 - 0.03}{0.0038} = 5.26.$$

This is so large a z-score that it's off the table; we conclude that the likelihood of this is essentially 0%, and therefore must conclude that $p \leq 0.03$ is incorrect.

11. $s^2_{\text{kings}} = 184.67; \quad s^2_{\text{pres}} = 38.37;$

$$\sigma_{\text{DDSM}} \cong \sqrt{\frac{184.67}{20} + \frac{38.37}{40}} = 3.19;$$

$$(\bar{x}_{\text{kings}} - \bar{x}_{\text{pres}}) \pm 1.96(3.19) = (40.4 - 54.8) \pm 6.25$$
$$\underset{\substack{\uparrow \\ 95\% \text{ confidence}}}{} = (-20.68, \ -8.17)$$

12. $\bar{p}_A = \frac{40}{100} = 0.40; \quad \bar{p}_B = \frac{60}{200} = 0.30;$

$$\sigma_{\text{DDSP}} \cong \sqrt{\frac{0.40(1 - 0.40)}{100} + \frac{(0.30)(1 - 0.30)}{200}}$$

$$= 0.0587$$

A 90% confidence interval for $(\bar{p}_A - \bar{p}_B)$ is

$$(0.40 - 0.30) \pm 1.645(0.0587) = 0.10 \pm 0.0966$$
$$\underset{\substack{\uparrow \\ 90\%}}{} = (0.0034, \ 0.1966).$$

Yes, A is better than B at the 90% level (since the interval does not contain 0 or any negative values).

ANSWERS TO
CHAPTER 3 QUIZ

1. confidence interval type

2. distribution; sample mean

3. distribution; sample statistic, or estimator

4. divided by 35

5. normal

6. fraction, or percentage

7. binomial

8. proportion

Hypothesis Testing

Terms we'll be learning about in this chapter

chi-square

contingency table

hypothesis

null hypothesis

one-tailed test

two-tailed test

4-1 INTRODUCTION

We are going to ask certain questions about population parameters, and we are going to call these questions hypotheses. A **hypothesis** is an assumption to be tested (to determine if it is true or false).

Suppose I take a coin out of my pocket and flip it 10 times. And suppose it comes up heads every time. Can this be a fair coin? I wonder. Well, how likely is it that a fair coin would come up heads 10 times out of 10? This is a statistics question, for it has to do with a certain distribution (the binomial distribution with $p = \frac{1}{2}$) and how likely a certain outcome from that distribution is. Recall formula (9) of Chapter 2 (page 74) for the binomial distribution with $p = \frac{1}{2}$; we see that the frequency of occurrence of the outcome 10 heads out of 10 tosses is

$$\binom{10}{10}\left(\frac{1}{2}\right)^{10} = 0.001 \quad \text{(actually, 0.0009766).}$$

So we are faced with two possible conclusions: either the coin is **fair** (that is, $p = \frac{1}{2}$) and what I observed was a *rare event* (in this case, less than 1 time in 1000), or the coin is **biased** (that is, $p \neq \frac{1}{2}$). "$P < 0.001$" means "the probability of this event is less than 0.001." Do not confuse P with p, the probability of heads in one toss.

"Sorry, Fred, it's heads again."

"I can't understand how I've lost so many times in a row."

This is all statistics can do for us. We must then decide which conclusion to choose. Most people, of course, would assume that the coin is biased, rather than assume that I've observed a rare event (since *by definition* rare events don't happen very often). And when we **reject the hypothesis** that $p = \frac{1}{2}$, we reject it at the 0.001 **level of significance**, which means that to not reject the hypothesis would have required acknowledging a rare ($P \leq 0.001$) event. In statistics literature and elsewhere, significance levels are frequently attached to conclusions, usually in the form $P \leq 0.001$ or $P < 0.01$ or $P < 0.05$, which just means the probability of the rare event is less than or equal to 0.001 or 0.01 or 0.05.

Notice that although the probability of observing 10 heads out of 10 flips of a fair coin is 0.0009766, we state that $P < 0.001$. This kind of simplification is always done in hypothesis testing, and the usual levels are 0.001, 0.01, 0.05, and 0.10. So if the probability had been 0.023, we would say $P < 0.05$ because 0.05 is the *smallest* of the numbers 0.001, 0.01, 0.05, and 0.10 for which the statement $P < $ _____ is true. The custom is: after computing the true probability (0.0009766 or 0.023 or whatever), go to the next largest number of 0.001, 0.005, 0.01, 0.05, and 0.10 and use that one. Actually, my own preference would be to use the exact computed probability, as this communicates more information, but as members of the statistical community, we have to conform to "what's done."

TESTING BINOMIAL PROPORTIONS: TWO-SIDED ALTERNATIVES

Let's continue to work with the binomial distribution as our example. Suppose we flip a different coin 10 times and get 2 heads out of 10 tosses. Can we conclude that the coin is *biased* (that is, $p \neq \frac{1}{2}$)? To study whether the coin is biased, we need to examine the *alternative*—that the coin is *not biased*. The reason for this is as follows. To say that the coin is biased includes all sorts of possibilities: $p = 0.2$ (a reasonable possibility, as we observed 2 heads out of 10), or $p = 0.99$ (an unreasonable possibility), and so forth. It would be difficult to develop a test that would contain both a reasonable and an unreasonable possibility at the same time, because we would expect the answer to be "yes" in the reasonable case and "no" in the unreasonable case. Consequently, the hypothesis we set up for testing must have the property that *if the hypothesis is true, we know what to expect.* (This is a very important concept.) So, even though we wish to study whether *or not* $p = \frac{1}{2}$, we directly test the question $p = \frac{1}{2}$ rather than $p \neq \frac{1}{2}$, because when $p = \frac{1}{2}$, *we know what to expect*, whereas when $p \neq \frac{1}{2}$, *we don't know what to expect*. We call the hypothesis that we can test directly the **null hypothesis**. The "other" hypothesis (that the null hypothesis is false) is called the **alternative hypothesis**. Continuing, if $p = \frac{1}{2}$ is true, we know the distribution of the number of heads out of 10 tosses of the coin. It's the binomial distribution with $p = \frac{1}{2}$, $n = 10$. The likelihood of observing 2 heads out of 10 tosses, by formula (9), Chapter 2, is _____ . (Take the time to use your calculator to obtain the value. Did you get 0.044?) Does that mean we would reject the hypothesis $p = \frac{1}{2}$ at the 0.05 level? Not exactly, for the following two reasons.

First, if we consider 2 heads out of 10 tosses "unusual," how would we regard 1 head out of 10 tosses, or 0 heads out of 10 tosses? Everyone would agree that if 2 is "unusual," then 1 or 0 is also unusual. For that reason, rather than consider the likelihood of observing *exactly* 2 heads out of 10 tosses, we consider the likelihood of observing 2 heads or *fewer* out of 10 tosses. (If you can't see this, imagine tossing a coin 1000 times. The number of possible heads you could observe varies from 0 to 1000, or 1001 possible values. There are so many different numbers of heads possible that any single number of heads, say exactly 500, is very unlikely by itself.) So we can't consider the likelihood of an *individual* outcome as likely or unlikely, because in a situation like this, any outcome by itself is rather unlikely. We therefore consider only aggregates or collections of outcomes and their likelihood. If we consider the possibility that 2 heads out of 10 is unlikely, we ought therefore to consider 2 heads *or fewer* as an aggregate. In other words, (the probability of 0 out of 10) + (the probability of 1 out of 10) + (the probability of 2 out of 10). By formula (9), the probability of

$$0 \text{ out of } 10 = \binom{10}{0}\left(\frac{1}{2}\right)^{10} = 0.0009766$$

$$1 \text{ out of } 10 = \binom{10}{1}\left(\frac{1}{2}\right)^{10} = 0.0097656$$

$$2 \text{ out of } 10 = \binom{10}{2}\left(\frac{1}{2}\right)^{10} = 0.0439450$$

$$= 0.0547$$

So, the probability of 2 or fewer out of 10 is 0.0547.

The second reason we would not reject 2 out of 10 as unlikely is that we have to consider how we would react to 8 out of 10 heads instead of 2 out of 10. If we have no prior reason to consider values of p less than $\frac{1}{2}$ preferentially over values of p greater than $\frac{1}{2}$, then observing 8 out of 10 heads is just as much cause for concern that p is or is not $\frac{1}{2}$ as observing 2 out of 10 heads. Thus, we are testing $p = \frac{1}{2}$ against the **two-sided alternative** $p \neq \frac{1}{2}$, which means that we are considering both exceptionally large and exceptionally small values for the number of heads *all together*. Since the distribution of outcomes for the binomial distribution is *symmetric* when $p = \frac{1}{2}$, the probability of 8 heads out of 10 tosses is the same as the probability of 2 heads out of 10 tosses; the probability of 9 out of 10 is the same as the probability of 1 out of 10; and the probability of 10 out of 10 is the same as the probability of 0 out of 10. Thus, when we observe 2 out of 10 heads and ask if this is unlikely, we really consider both large and small outcomes at the same time: namely, 0, 1, or 2 heads, together with 8, 9, or 10 heads. The total probability of this aggregate of outcomes is $2(0.055) = 0.11$. Unless we consider events that happen 11% of the time as rare, we do not reject the hypothesis $p = \frac{1}{2}$.

Let's review what we've just covered. If we observe 2 heads out of 10 flips and ask, "Is the coin biased?" we reverse the question and ask, "Is the coin fair?" because *that* question leads to knowledge of the distribution (where asking "Is the coin biased?" doesn't). Then we ask for the probability not of *exactly* 2 heads, but of 0, 1, or 2 heads, together with 8, 9, or 10 heads, as these outcomes are all as extreme or more so than 2 heads. The total probability of these outcomes is, by formula (9) of Chapter 2, $0.001 + 0.010 + 0.044 + 0.044 + 0.010 + 0.001 = 0.11$, and we do not reject this, since we do not consider events that happen 11% of the time as rare events.

A frequent question is "How do I decide what *rare* is? Is 5% rare? Or only 0.1%?" The answer is, it depends on the situation and how important it is *to you* not to be wrong (in all statistical inference, you *must* be wrong some fraction of the time). If you are testing tires, 5% error may be acceptable. But if you are looking at birth control devices, you probably won't accept more than a 0.01 or 0.001 error. What's really involved here is the relative costs of making two kinds of errors. In a cancer cure, the two possible kinds of error are (1) deciding that a new drug is effective when it isn't, and (2) deciding that a new drug is not effective when it is. We have to balance the personal costs of these two types of errors in deciding where to set our level of significance. (See Section 4-2 for further discussion.)

TESTING BINOMIAL PROPORTIONS: ONE-SIDED ALTERNATIVES

Suppose the Thayer Drug Company is in the process of developing a new headache cure. Suppose further that the company requires that the drug have *at least* a 60% cure rate; that is, it must cure at least 60% of the headaches it treats, within one hour, say. This is a binomial population (cured and not cured; with cured = 1, not cured = 0) with p = the percentage cured. An experiment is carried out on 100 subjects with headaches, and 69 are cured. Is the drug *at least* 60% effective?

As I said earlier, to test a hypothesis, we must have a testable hypothesis—that is, one for which if true, we know the distribution of the so-called test statistic. The conclusion Thayer wants to test is $p \geq 0.60$. The alternative hypothesis is $p < 0.60$. Consider for the moment one single value p could take—say, $p = 0.50$—*within* the aggregate alternative hypothesis $p < 0.60$. If we can reject $p = 0.50$ on the basis of 69 cures out of 100, then we can also reject *any* $p < 0.50$. That's because if we find that observing 69 cures out of 100 is unlikely with $p = 0.50$, then it's *even more unlikely* that we observe 69 cures out of 100 with $p = 0.49$. And for that reason, if we can reject $p = 0.50$, then we can reject $p = 0.49$. Also, if we can reject $p = 0.50$, then we can reject *all* $p < 0.50$. Thus, to reject $p < 0.60$, we test $p = 0.60$. And if we can reject $p = 0.60$ on the basis of 69 cures out of 100, then we can reject *all* $p < 0.60$. The company can then conclude $p > 0.60$. (Notice how we started out to test $p \geq 0.60$ and actually concluded the stronger result, $p > 0.60$.)

Now it's time to actually test $p \geq 0.60$ with 69 cures out of 100. We test (for the reasons discussed above) the null hypothesis $p = 0.60$. As I said in Section 3-2, the binomial distribution is approximately normal when $n\bar{p}$ and $n(1 - \bar{p})$ are both greater than 5. If $p = 0.60$ and $n = 100$, both these criteria are satisfied, and we know that the binomial distribution is approximately normal with mean $np = 60$ and variance $np(1 - p) = 24$.

The probability we wish to compute is the probability of 69 *or more* successes out of 100 if $p = 0.60$. Using the normal approximation to the binomial, we approximate the shaded area of Figure 4-1, that is, the portion of the normal to the right of 68.5 rather than 69. This is because the bar over the class 69 really extends from 68.5 to 69.5. So to approximate the area, we must start at 68.5. This is called the **continuity correction** of the normal approximation to the binomial. We compute the z-score of 68.5:

$$z = \frac{68.5 - 60}{\sqrt{24}} = 1.74.$$

Depending on whether we want 69 or more or 69 or less determines whether we compute

$$z = \frac{(69 - 0.5) - 60}{\sqrt{24}} \quad \text{or} \quad z = \frac{(69 + 0.5) - 60}{\sqrt{24}}.$$

When you make the continuity correction, draw a picture like the one in Figure 4-1 to see whether to add or subtract 0.5.

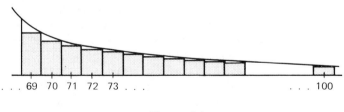

. . . 69 70 71 72 73 100

Figure 4-1

Now that we've made the continuity correction, the value 69 corresponds to a z-score of

$$z = \frac{68.5 - 60}{\sqrt{24}} = 1.74$$

(the standard deviation is $\sqrt{24}$). The probability of a z-score as large or larger than 1.74 is

$$0.5000 - \text{Prob}(z\text{-score is less than 1.74 but greater than 0}) = 0.5000 - 0.4591,$$

and looking up 1.74 in Table 1 at the back of the book,

$$= 0.0409,$$

and so we reject the hypothesis $p = 0.60$ at the 0.05 level of significance ($P < 0.05$) and conclude $p > 0.60$. *Note that this is a* **one-sided alternative**. Note also that we used only *one tail* of the normal distribution because we were testing a one-sided alternative. This is frequently called "making a **one-tailed test**."

To summarize, to make a one-tailed test at the 0.05 level, you need to find 5% *or less* of the distribution in *one* of the tails. To make a **two-tailed test** at the 0.05 level, you need to find 2½% or less in each tail (to obtain 5% or less in both tails).

EXERCISE 1   Use the normal approximation to the binomial distribution to compute the probability

a. of flipping a *fair coin* ($p = ?$) 100 times and getting *more than* 55 heads

_____

_____

b. of getting 40 *or more* heads. Remember the continuity correction. (*Hint*: Draw a picture as we did in Figure 4-1.)

_____

_____

EXERCISE 2   Do as requested in Exercise 1, but this time the probability of heads is 0.40.

a. _____

_____

b. _____

_____

EXERCISE 3   The Widget Company claims that their defective rate (percentage of widgets manufactured that are defective) is 5% or less. A sample of 300 widgets is tested, and 30 are found to be defective. Test the company's claim at the 0.05 level of significance. (This is a binomial problem, one-tailed, with continuity correction.)

_____

_____

_____

4-2 TWO TYPES OF ERROR

It's cloudy this morning, and I wonder whether or not I should take my umbrella. There are two possible kinds of error: either I take my umbrella and it doesn't rain, or I don't take it and it does rain. We can't prevent these errors, but we can control them by weighing the *relative costs* of one type versus the other. Most people would say that carting around an unused umbrella is less of a cost than getting soaked in a downpour, but each individual has to decide.

In deciding whether to accept a hypothesis, we are always subject to the same two kinds of error: we can accept the hypothesis as true, when it isn't, or we can reject the hypothesis as false, when it is true (Figure 4-2). In the Thayer Drug Company example, the company weighs the relative costs of the two kinds of error: the error of deciding that the drug is effective when it isn't (which opens the company up to future lawsuits, bad publicity, and so on) and the error of deciding that the drug is not effective when it is (which costs the company all that business as a result of not marketing the drug). So the company balances these two costs and sets a

Umbrella

No umbrella

Rain   No rain

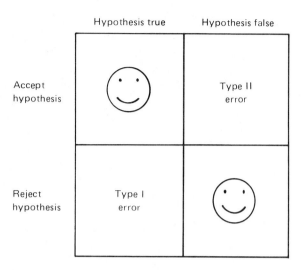

Figure 4-2   Two types of error.

cutoff point, called the **critical value**, which is a threshold above (or below) which the company decides the drug is (or is not) effective. If, instead, the company wishes to reduce the probability of *both* kinds of error, it must then increase the sample size of the experiment. This is typical when facing two kinds of errors in hypothesis testing and reducing them both.

EXERCISE 4   What are the two kinds of error associated with auto insurance? Make up some reasonable estimates of probabilities and costs associated with each.

_____

_____

_____

_____

_____

4-3 TESTS ABOUT MEANS: ONE-TAILED

Suppose a tire manufacturer claims that his tires get at least 40,000 miles on the average. This is the kind of claim we can test using hypothesis-testing techniques. We proceed just as we did with the Thayer Drug Company example of Section 4-1, except there we had a binomial population, and here we don't.

Suppose the manufacturer tests a sample of 35 of his tires and gets a sample mean of 40,900 and a sample variance $s^2 = 8,700,000$. Let's test his claim that the population mean is 40,000 or greater. To make a statistical test, we need a null hypothesis that leads to knowledge of the distribution. Suppose, just for the sake of argument, that we test $\mu = 39,000$. If $\mu = 39,000$, how likely is it that we would draw a sample of size 35 and observe a sample mean as large as 40,900 *or larger*? We know that to answer this we need to consider the DSM, which is approximately normal with mean (under the assumed hypothesis) $\mu = 39,000$ and variance $8,700,000/35$. Then $\sigma_{DSM} \cong \sqrt{8,700,000/35} = 498.57$.

Since we are starting with a claim that the mileage is *at least* 40,000, we have a *one-sided test* and calculate the probability of observing a value of 40,900, *or larger*, from a normal distribution with mean 39,000 and standard deviation 498.57. We compute a z-score:

$$z = \frac{40,900 - 39,000}{498.57} = 3.81.$$

There's no need to look that up in a normal table. The probability of such a z-score or larger is infinitesimal (the probability of a z-score as large or larger than 3 is already about 0.001).

Now here goes the "complicated" part of the argument, already contained in the drug company example. We can reject at, say, the 0.05 level the hypothesis $\mu = 39,000$, by virtue of the calculation we made above. But we can then also reject $\mu = 38,000$, because that value would lead to an *even larger* z-score. If you don't see this, think about it for a moment. The basic idea is simple. If 40,900 for a sample mean is unlikely when the population mean is 39,000, then 40,900 for a sample mean is *more unlikely* when the population mean is 38,000 (because it leads to a larger z-score). For this reason, if we can reject $\mu = 39,000$ against a one-sided alternative, based on an observed sample mean of 40,900, then we can reject 38,000; in fact we can reject all $\mu < 39,000$. Hence, if we want to reject *all* $\mu < 40,000$ (that's what the manufacturer wants), all we have to do is test $\mu = 40,000$ against the *one-sided alternative* (that is, with a one-tailed test) that $\mu > 40,000$. If we can reject $\mu = 40,000$ as a one-tailed test, then we can reject all $\mu < 40,000$. That's what we want; now let's proceed to make the test.

To test $\mu = 40,000$ against the one-sided alternative $\mu > 40,000$ at the 5% level, we compute a z-score:

$$z = \frac{40,900 - 40,000}{498.57} = 1.805.$$

From Table 1 (at the back of the book) the probability of $z \geqslant 1.805 = 0.0355$; since $0.0355 < 0.05$, we reject $\mu = 40,000$, and for the reasons we talked about, reject all $\mu \leqslant 40,000$. We conclude $\mu > 40,000$. Note that while the manufacturer claims that his tires get *at least* 40,000 miles on average, what we actually concluded was more: that his tires get *more than* 40,000 miles on average.

When we make a one-tailed test, say, at the 5% level, we concentrate all the 5% in the one tail: the z-score that is the critical value for a 5% tail of the normal distribution is the (critical) value 1.645. If we get a z-score of 1.645 or larger, we reject the null hypothesis at the 5% level. If we are making a two-tailed test, as we will in Section 4-4, then we put 2½% in *each* of the two tails for our 5%; then the 5% critical value for the normal distribution is ±1.96, because it can be either "large" or "small," and we have cause to reject the null hypothesis in either case. In the one-tailed test, we reject only if we find ourselves on the one side for rejection, as in the tire example above.

In summary, then, the 5% critical value for the normal distribution is ±1.96 if two-tailed and 1.645 if one-tailed (see Figure 4-3).

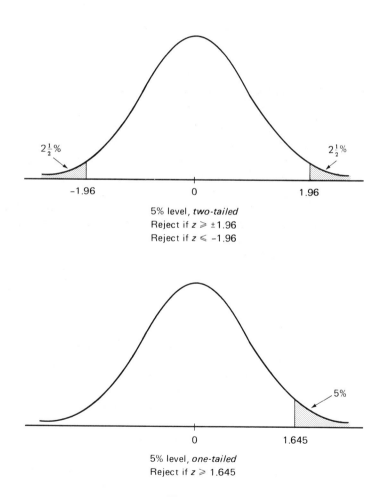

Figure 4-3

EXERCISE 5

a. The fire department claims that it responds to emergencies in an average of 5 minutes *or less*. The next 40 service calls are studied. They reveal an average response time of 5.8 minutes, with a sample standard deviation of 3 minutes. Test the fire department's claim at the 0.05 level of significance. (Is this one- or two-tailed?)

_____

_____

_____

b. What if the average response for the 40 calls is 5.7 minutes?

_____

_____

_____

4-4 TESTS ABOUT MEANS: TWO-TAILED

Suppose the City Police claim that their average response time to emergency calls is 10 minutes. We wish to test this claim and will reject it if the figure we get is either too large or too small. In this case we will be making a two-tailed test. The null hypothesis (the one that leads to knowledge of the distribution of the statistic) is $\mu = 10.0$ against the alternative hypothesis $\mu \neq 10.0$. Suppose we observe 50 police calls and find that the average response time of these 50 calls is 9.6 minutes, with a sample standard deviation of 1.7 minutes. Since n is greater than 30, the DSM is approximately normal, and under the null hypothesis $\mu = 10.0$, the mean of the DSM is 10.0. We compute a z-score:

$$z = \frac{9.6 - 10.0}{1.7/\sqrt{50}} = -1.66.$$

Since $z = -1.66$ is not greater than $+1.96$ or smaller than -1.96, we cannot reject the hypothesis $\mu = 10.0$ at the 5% level of significance (see Figure 4-3). We conclude that we have no basis for rejecting the police department's claim.

Now you know enough to understand some of the "jargon" of statisticians and scientists. Frequently you'll overhear phrases like "this was a 2σ event," or "this was a 3σ event." What this means is that (in whatever the circumstance happens to be) assuming the null hypothesis is true, the particular event observed has a z-score of 2 (or greater). That's what a z-score measures (sigmas). We know that a 2σ event (z-score of 2 or greater) has a probability of occurrence less than 5% (actually, 4%), and that a z-score of 3 ("3σ event") has a probability of less than 1% (actually, 0.3%). So 2σ events occur less than 5 times out of 100, and 3σ events occur less than 3 times in 1000 (two-sided, or two-tailed).

EXERCISE 6   The National Electric Company and the union want to agree on an average time to assemble one toaster. They tentatively agree on 37 minutes per toaster. To test this, they keep careful measurements on the time it takes to assemble the next 50 toasters. It turns out that the 50 times have a sample mean of 36.3 minutes and a sample standard deviation of 2.8 minutes. Test (two-tailed) whether or not the data support the claim that $\mu = 37$, at the 0.05 level of significance.

4-5 TESTING DIFFERENCES BETWEEN MEANS (INDEPENDENT SAMPLES)

Do high school students who watch several hours of television each day perform as well (as measured by grade-point average) as students who watch little or no television? Do apprenticeship-trained mechanics perform as well (as measured by the length of time taken to tune an automobile engine) as factory-trained mechanics? Do rats fed a diet of white bread grow to the same size as rats fed a balanced diet? These questions, as well as dozens you can think of yourself, all ask if the means of two populations are the same or different. We can use the machinery of hypothesis testing to answer these questions once we put the questions into the proper framework.

"I'll have a BLT on whole wheat, hold the mayo."

Suppose we are given two populations, population A and population B. We are also given samples A and B, with sample means $\bar{x}_A$ and $\bar{x}_B$, respectively. We wish to know, on the basis of these samples and sample means, whether or not we can reject the hypothesis $\mu_A = \mu_B$ against the alternative hypothesis $\mu_A \neq \mu_B$. To answer this, we make use of the following facts:

1. The distribution of differences between sample means, $\bar{x}_A - \bar{x}_B$, is approximately normally distributed if the sample sizes n_A and n_B are each at least 30.

2. The mean of the distribution is the difference between the population means, $\mu_A - \mu_B$.

3. The variance of the distribution is the *sum* of the variances of the two DSMs of $\bar{x}_A$ and $\bar{x}_B$.

Now suppose we have a sample of size 40 from population A, with sample mean $\bar{x}_A = 50$ and sample variance $s_A^2 = 10$, and a sample of size 35 from population B, with sample mean $\bar{x}_B = 48$ and sample variance $s_B^2 = 16$. And we assume that the selection of elements for one sample has nothing to do with the selection of elements for the other sample. This is what we mean by **independent samples**. We compute a z-score for the differences between means, noting that, from item 3 above, the variance of the distribution of $\bar{x}_A - \bar{x}_B$ is $10/40 + 16/35$. Thus,

$$z = \frac{\bar{x}_A - \bar{x}_B}{\sqrt{\dfrac{s_A^2}{40} + \dfrac{s_B^2}{35}}} = \frac{50 - 48}{\sqrt{\dfrac{10}{40} + \dfrac{16}{35}}} = 2.38.$$

For a two-tailed test, this z-score is significant at the 0.05 level ($P < 0.05$), since z exceeds 1.96, but not at the 0.01 level, since z does not exceed 2.58. Thus we can say that the population means are different at the 0.05 level of significance.

If we wished to test the claim that the mean of population A is *greater* than the mean of population B, we would compute the same z-score, but make a one-tailed test. For the example just given, the z-score 2.38 would be significant at the 0.01 level ($P < 0.01$), since it exceeds 2.33.

The formula for computing the z-score for the difference between two sample means, given large sample sizes, is

$$z = \frac{\bar{x}_A - \bar{x}_B}{\sqrt{\dfrac{s_A^2}{n_A} + \dfrac{s_B^2}{n_B}}} . \tag{1}$$

If one of the sample sizes is less than 30, then we use the same ideas of hypothesis testing, but use the small-sample t-distribution (see Chapter 5 for details).

EXERCISE 7   A psychologist wishes to test whether there is any difference in puzzle-solving abilities between boys and girls. Forty boys chosen at random took an average of 5 minutes to solve a puzzle, with a sample standard deviation of 5.3 minutes. Forty-five girls took an average of 4.8 minutes to solve the same puzzle, with a sample standard deviation of 4.6 minutes.

   a. Test the hypothesis that the girls are *better* puzzle solvers than the boys. (Is this one-tailed or two-tailed?)

   _____

   b. Test the hypothesis that there is *no difference* in puzzling-solving ability between the boys and the girls. (One-tailed or two-tailed?)

   _____

4-6 CONTINGENCY TABLES

Is income level related to political affiliation? Are women more likely to be better students in this class than men? Is a woman's hair color related to her social success (or, do blondes really have more fun)? These questions are all of the same type: they all ask if *population proportions are the same or different in two or more populations*. These questions can all be answered by formulation of a test statistic whose distribution will be known if the proportions are all the same.

Let's look at an example. In a survey, 200 people are asked to rate themselves as Republican, Democrat, or other, and also to rate themselves as low income, middle income, or high income. Each of the 200 surveyed thus falls into one of the nine cells shown in Figure 4-4. Thus, we tally the 200 responses and enter the

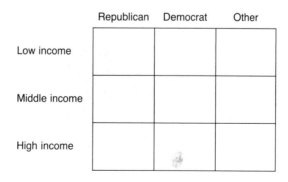

Figure 4-4

counts in the nine cells. These **count data**, as they are called, together with the nine cells arranged in a 3 × 3 array, make up what is called a 3 × 3 **contingency table**. Let's suppose the numbers are as shown in Figure 4-5. Is political affiliation related to income level? (Are high-income people less likely to be Democrats than low-income people?)

	Republican	Democrat	Other
Low income	9	35	8
Middle income	8	27	35
High income	40	30	8

Figure 4-5

As before, it's difficult to test this question directly, because it includes so many *possible* relationships. We can, however, test the opposite hypothesis directly (the hypothesis of *no* relationship between the column variable and the row variable),

because if that hypothesis is true (if there is no relationship between income and political affiliation), then *we know what to expect in the way of outcomes.* Observe that there are $9 + 8 + 40 = 57$ Republicans, 92 Democrats, and 51 others. And there are 52 low incomes, 70 middle incomes, and 78 high incomes (see Figure 4-6).

	Republican	Democrat	Other	Totals
Low income	9	35	8	52
Middle income	8	27	35	70
High income	40	30	8	78
Totals	57	92	51	200

Figure 4-6

If there is no relationship between income and political affiliation, then low-income individuals, for example, should be Republicans, Democrats, and others roughly in the proportions $57:92:51$. Does this mean that if we fail to see *exactly* these proportions, the hypothesis of no relationship is to be rejected? Certainly not, but these proportions at least tell us what numbers to expect to see in the various cells: of the 52 low-income individuals, $57/200$ of them are expected to be Republicans, or

$$\frac{52 \times 57}{200} = 14.82;$$

$92/200$ of the 52 low-income individuals are expected to be Democrats, or

$$\frac{52 \times 92}{200} = 23.92;$$

and finally, $51/200$ of the 52 low-income individuals are expected to be others, or

$$\frac{52 \times 51}{200} = 13.26.$$

So, while we *observe* in the first row of our contingency table

9	35	8

we *expect* in the first row of our table

14.82	23.92	13.26

if the hypothesis of no relationship is true.

We repeat the same calculation for the other rows of the table. To streamline the computation, we observe that the number we *expect* in each cell is the product of the row and column totals for that cell divided by the total number surveyed. So, to compute the expected number of low-income Republicans, we multiply 52 by 57 and divide by 200, obtaining 14.82 (see Figure 4-7). We place these **expected numbers** in parentheses in the same cell along with the **observed numbers** (the counts).

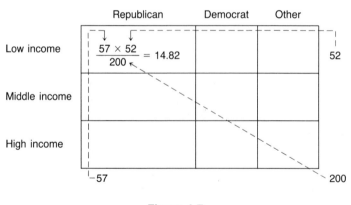

Figure 4-7

At this time, take your calculator and complete the table in Figure 4-8. (If you are unable to figure out how to do this, turn the page and look at Figure 4-9, where I've computed the expected numbers for you.)

Now we have the 9 numbers we *observe* and the 9 numbers we *expect* if the hypothesis of no relationship is true. But we still need to calculate some *one statis-*

	Republican	Democrat	Other	Totals
Low income	9 (14.82)	35 (23.92)	8 (13.26)	52
Middle income	8 ( )	27 ( )	35 ( )	70
High income	40 ( )	30 ( )	8 ( )	78
Totals	57	92	51	200

Figure 4-8

tic that measures how close the 9 observed numbers are to the expected numbers. To do this, we take the difference between the observed number and the expected number, and square it. Then we divide by the expected number. Do this for each of the 9 entries, and add up the 9 numbers you obtain. This value is the statistic called **chi-square** and is denoted

$$\chi^2 = \sum_{i=1}^{9} \frac{(o_i - e_i)^2}{e_i} ,$$

where $\sum_{i=1}^{9}$ means "add up 9 things labeled 1, 2, 3, . . . , 9," o_i means the ith ob-
served cell entry, and e_i means the corresponding cell expected value. It makes no difference which cell corresponds to 1, 2, and so on. A moment's reflection will tell you that the closer the observed values are to the expected values, the smaller the value of the statistic. And the larger the value of the statistic, the more deviation there is of the observed values from the expected values.

If the hypothesis of no relationship (null hypothesis) is correct, it is possible mathematically to compute the distribution of the statistic chi-square; and its distribution is called the **chi-square distribution**. There isn't just one chi-square dis-
tribution. There is a whole family of them, indexed by "degrees of freedom." But we'll refer to them all as the chi-square distribution; this distribution appears in Table 4 at the back of the book. The number of degrees of freedom is 1 less than the number of rows, times 1 less than the number of columns, or in this case we have $(3 - 1) \times (3 - 1) = 2 \times 2 = 4$ degrees of freedom.

In general, we will have m categories for one variable (m rows) and n categories for the other variable (n columns), as in Figure 4-10, page 154.

	Republican	Democrat	Other	Totals
Low income	9 (14.82)	35 (23.92)	8 (13.26)	52
Middle income	8 (19.95)	27 (32.20)	35 (17.85)	70
High income	40 (22.23)	30 (35.88)	8 (19.89)	78
Totals	57	92	51	200

Figure 4-9

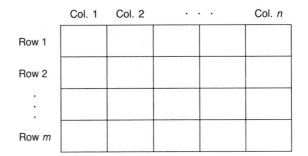

Figure 4-10   An $m \times n$ contingency table has $m \cdot n$ cells. Compute

$$\chi^2 = \sum_{i=1}^{m \cdot n} \frac{(o_i - e_i)^2}{e_i}$$

with $(m - 1) \cdot (n - 1)$ degrees of freedom.

Now that we know what the distribution of the statistic is, assuming the null hypothesis, we treat this just like any other hypothesis-testing question. We calculate the value of the statistic under the null hypothesis. (We have to assume the null hypothesis; otherwise we don't know the distribution of chi-square.) Then we see if the value is "unreasonably" large. If it is, either we have a "rare event," or the hypothesis we have been assuming (the hypothesis of no relationship between political affiliation and income level—the null hypothesis) is false. Since we don't accept rare events, we reject the null hypothesis if the significance level is low enough.

We calculate for our example above, using the observed and expected entries in Figure 4-8 or 4-9:

$$\frac{(9 - 14.82)^2}{14.82} + \frac{(35 - 23.92)^2}{23.92} + \frac{(8 - 13.26)^2}{13.26}$$

$$+ \frac{(8 - 19.95)^2}{19.95} + \frac{(27 - 32.2)^2}{32.2} + \frac{(35 - 17.85)^2}{17.85}$$

$$+ \frac{(40 - 22.23)^2}{22.23} + \frac{(30 - 35.88)^2}{35.88} + \frac{(8 - 19.89)^2}{19.89}$$

$$= 56.26.$$

Thus, the value of the chi-square statistic is 56.26.

Now we look in the table of critical values of the chi-square distribution, Table 4 (in the back of the book). Notice that each row of the table is indexed by d.f., which stands for "degrees of freedom." We know that we must look under 4 d.f., which is the fourth row of the table. Figure 4-11 shows the way the right half of row

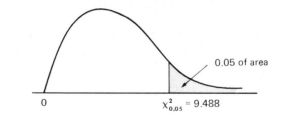

d.f.	$\chi^2_{0.05}$	$\chi^2_{0.025}$	$\chi^2_{0.01}$	$\chi^2_{0.005}$	d.f.
.					.
.					.
.					.
4	9.488	11.143	13.277	14.860	4
.					.
.					.
.					.

Figure 4-11

4 looks in the table. Don't worry about the left half for now. The way we read this table is as follows: the value 9.488 is the 0.05, or 5%, critical value of chi-square, 4 d.f. That means that 5% of the distribution lies to the *right* of (that is, is larger than) the value 9.488. And 2½% of the distribution lies to the right of 11.143. Similarly, 0.01, or 1%, of the distribution lies to the right of 13.277 and 0.005, or 0.5%, of the distribution lies to the right of 14.860. Our value, 56.26, therefore lies in the ½%, or 0.005, tail of the distribution because it is larger than (that is, to the *right* of) the value 14.860. That tells us that the likelihood of observing 56.26 by chance, if the null hypothesis is true, is less than 0.005, an exceedingly rare event. Therefore, we reject the null hypothesis (the hypothesis of no relationship) at the 0.005 level of significance ($P < 0.005$) and conclude that based on these data, there is a relationship between political affiliation and income level.

A WORD OF CAUTION

Do not compute chi-square for contingency tables when any of the expected cell frequencies is less than 5. The reason for this is that one single large deviation can make the value of chi-square "significant" if the number you divide by is small. The divisors e_i in the statistic are the expected values for the cells. If any one of the expected values is less than 5, combine two or more of the rows or columns, so that the resulting entry has an expected value greater than 5. If, in our example above, the expected value for the upper right cell ("Low income" and "Other") were smaller than 5, then we would combine "Other" with "Republican" and make our test on the resulting 3×2 array shown in Figure 4-12.

	Republican and Others	Democrat
Low income	17	35
Middle income	43	27
High income	48	30

Figure 4-12

EXERCISE 8   Calculate chi-square for the 3×2 array in the figure below.

	Republican and Others	Democrat
Low income	17 ( )	35 ( )
Middle income	43 ( )	27 ( )
High income	48 ( )	30 ( )

EXERCISE 9   What is the significance level of chi-square in Exercise 8?

4-7 MEASURING GOODNESS OF FIT

The notion of matching observed values with expected values in count data can also be used to test a hypothesis as to whether or not certain data come from (that is, are *fit by*) some given distribution. According to the hypothesis, the data we are given come from this distribution. If we form intervals or classes from within the hypothesized distribution, we know according to the distribution how many of each class we can expect. We use chi-square to measure how likely or unlikely the observed event is. This is called **measuring the goodness of fit**.

FITTING TO THE UNIFORM DISTRIBUTION

Suppose we roll a six-sided die 120 times. It comes up 1, 2, 3, 4, 5, 6 the following number of times:

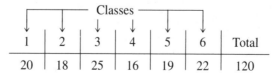

	1	2	3	4	5	6	Total
	20	18	25	16	19	22	120

If the die is "fair," we expect that it is equally likely that any face will come up. Therefore, we expect 1/6 of the rolls to be "1," 1/6 to be "2," and so on. Thus, in 120 rolls, we *expect* 20 1s, 20 2s, and so on. When all the outcomes are equally likely, we call the distribution a **uniform distribution**.

	1	2	3	4	5	6	Total
Observed:	20	18	25	16	19	22	120
Expected:	(20)	(20)	(20)	(20)	(20)	(20)	

We calculate chi-square:

$$\chi^2 = \sum_{i=1}^{6} \frac{(o_i - e_i)^2}{e_i} = \frac{(20-20)^2}{20} + \frac{(18-20)^2}{20} + \frac{(25-20)^2}{20}$$

$$+ \frac{(16-20)^2}{20} + \frac{(19-20)^2}{20} + \frac{(22-20)^2}{20}$$

$$= 0 + 0.2 + 1.25 + 0.8 + 0.05 + 0.2$$

$$= 2.50.$$

To calculate degrees of freedom, we do not take 1 less than the number of rows, because there is only 1 row; we take 1 less than the number of cells, or entries: $6 - 1 = 5$. We look up the 5% critical value for chi-square in Table 4, $\chi^2_{0.05}$, 5 d.f., and observe 11.07. Since our value of the statistic, 2.50, does not exceed the 5% critical value for chi-square, we conclude that the observation is not a "rare event," and we do not reject the null hypothesis (which in this case is that the six sides of the die are equally likely). We conclude that the die is fair.

FITTING TO ANY DISTRIBUTION

The same ideas used to test the uniform distribution (such as die rolls) can be used to test whether given data come from *any* specified distribution. First we use the data to estimate the parameters (if any) of the distribution. Then we group the data into classes, making the classes large enough so that if the data really come from the specified distribution, the expected number in each class will be at least 5 (so we can apply chi-square). To determine the expected number in a class, we need to calculate the area under the curve of the given distribution that that class contains, which is the relative frequency or proportion of that class, and multiply that proportion by the sample size. This is the *expected number* for that class. Now we have a collection of classes, expected numbers (calculated from the distribution), and observed numbers (counts from the data). We compute chi-square for these data and look up chi-square critical values for $(n - k - 1)$ d.f., where n is the sample size and k is the number of parameters in the distribution. (For example, if we were fitting to a normal, we would estimate the *mean* of the normal from the mean of the sample, and the *variance* of the normal from the variance of the sample. Thus, in this case, $k = 2$, and we use $n - 3$ d.f. In the case of the uniform distribution, the number of parameters is 0, and so $n - k - 1 = n - 1$ d.f.)

Many populations in real life are observed to be normally distributed. That's how the normal distribution was first discovered. As an example, let's test the hypothesis that the annual rainfall in Philadelphia (from Chapter 1) comes from a normal distribution. (These data are on page 6.) First we compute $\bar{x}$ and s^2: $\bar{x} = 42.06$ and $s^2 = 40.425$. Then we convert the tallies and histogram of Figure 1-7 (page 7) to a normal distribution with $\mu = 42.06$ and $\sigma^2 = 40.425$, and then we convert the histogram intervals to z-scores (see Figure 4-13). We then compute the areas under the normal distribution (Figure 4-14) whose boundaries are the class boundaries of the rainfall histogram:

$$(-2.68 \text{ to } -1.90): \quad \text{Prob}(-2.68 \leq z \leq -1.90) = 0.025$$

$$(-1.90 \text{ to } -1.13): \quad \text{Prob}(-1.90 \leq z \leq -1.13) = 0.1005$$

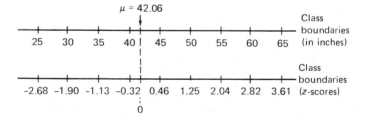

Figure 4-13

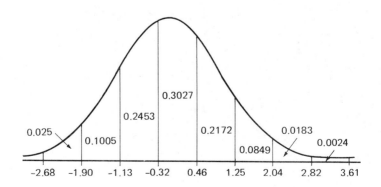

Figure 4-14

(−1.13 to −0.32): Prob(−1.13 ⩽ z ⩽ −0.32) = 0.2453

(−0.32 to +0.46): Prob(−0.32 ⩽ z ⩽ +0.46) = 0.3027

(0.46 to 1.25): Prob(0.46 ⩽ z ⩽ 1.25) = 0.2172

(1.25 to 2.04): Prob(1.25 ⩽ z ⩽ 2.04) = 0.0849

(2.04 to 2.82): Prob(2.04 ⩽ z ⩽ 2.82) = 0.0183

(2.82 to 3.61): Prob(2.82 ⩽ z) = 0.0024

Now we compute for a sample of $n = 120$ from this normal distribution how many entries we *expect* in each interval. As usual, if there are any expected values less than 5, we will pool cells.

(25–30 inches) = (−2.68 to −1.90)
→ (0.025)(120) = 3

(30–35 inches) = (−1.90 to −1.13)
→ (0.1005)(120) = 12.06

⎫ Pool

(35–40 inches) = (−1.13 to −0.32)
→ (0.2453)(120) = 29.44

(40–45 inches) = (−0.32 to +0.46)
→ (0.3027)(120) = 36.32

(45–50 inches) = (+0.46 to 1.25)
→ (0.2172)(120) = 26.06

(50–55 inches) = (1.25 to 2.04)
→ (0.0849)(120) = 10.19

(55–60 inches) = (2.04 to 2.82)
→ (0.0183)(120) = 2.20

⎫ Pool

(60–65 inches) = (2.84 to 3.61)
→ (0.0024)(120) = 0.29

Assuming a normal distribution, we *expect* and *observe* as follows:

Class	Expect (from normal)	Observe (actual count, p. 6)
25–35	15.06	17
35–40	29.44	31
40–45	36.32	31
45–50	26.06	29
50–65	12.68	12

Now we compute χ^2:

$$\chi^2 = \frac{(17 - 15.06)^2}{15.06} + \frac{(31 - 29.44)^2}{29.44} + \frac{(31 - 36.32)^2}{36.32}$$

$$+ \frac{(29 - 26.06)^2}{26.06} + \frac{(12 - 12.68)^2}{12.68}$$

$$= 1.48.$$

We have $n = 5$ cells; we estimated $k = 2$ parameters (μ and σ) of the normal, so we have $n - k - 1 = 5 - 2 - 1 = 2$ d.f. We note that $\chi^2 = 1.51$ is smaller than any critical value of χ^2, 2 d.f. So we cannot reject the hypothesis that these data do not come from a normal distribution. (That is, these data are compatible with a normal distribution, $\mu = 42.06$, $\sigma = 6.358$.) So the rainfall data *are* fit by a normal distribution.

PROBLEM SET A

1. We draw a sample of size 50 from a population and observe a sample mean of 5.8 and a sample standard deviation of 3. Test the hypothesis that the population mean is 5.1 or less

   a. at the 0.10 level of significance
   b. at the 0.05 level of significance
   c. at the 0.01 level of significance

2. We draw a sample of size 60 from a population and observe a sample mean of 11 and a sample standard deviation of 2. Test the hypothesis that the population mean is 10.6 or less

   a. at the 0.10 level of significance
   b. at the 0.05 level of significance
   c. at the 0.01 level of significance

3. We draw a sample of size 45 from a population and observe a sample mean of 90 and a sample standard deviation of 4. Test the hypothesis that the population mean is 89 or less

   a. at the 0.10 level of significance
   b. at the 0.05 level of significance
   c. at the 0.01 level of significance

4. We draw a sample of size 30 from a population and observe a sample mean of 25 and a sample standard deviation of 2.2. Test the hypothesis that the population mean is 24 or less

   a. at the 0.10 level of significance
   b. at the 0.05 level of significance
   c. at the 0.01 level of significance

5. We draw a sample of size 80 from a population and observe a sample mean of 100.7 and a sample standard deviation of 2.1. Test the hypothesis that the population mean is 100 or less

   a. at the 0.10 level of significance
   b. at the 0.05 level of significance
   c. at the 0.01 level of significance

6. We draw a sample of size 1000 from a population and observe a sample mean of 100.1 and a sample standard deviation of 2.1. Test the hypothesis that the population mean is 100 or less

   a. at the 0.10 level of significance
   b. at the 0.05 level of significance
   c. at the 0.01 level of significance

7. We draw a sample of size 2000 from a population and observe a sample mean of 100.1 and a sample standard deviation of 2.1. Test the hypothesis that the population mean is 100 or less

   a. at the 0.10 level of significance
   b. at the 0.05 level of significance
   c. at the 0.01 level of significance

8. We draw a sample of size 3000 from a population and observe a sample mean of 100.1 and a sample standard deviation of 2.1. Test the hypothesis that the population mean is 100 or less

   a. at the 0.10 level of significance
   b. at the 0.05 level of significance
   c. at the 0.01 level of significance

PROBLEM SET B

1. We draw a sample of size 55 from a population and observe a sample mean of 6.3 and a sample standard deviation of 3. Test the hypothesis that the population mean = 5.3

   a. at the 0.10 level of significance
   b. at the 0.05 level of significance
   c. at the 0.01 level of significance

2. We draw a sample of size 80 from a population and observe a sample mean of 11 and a sample standard deviation of 2. Test the hypothesis that the population mean = 10.6

   a. at the 0.10 level of significance
   b. at the 0.05 level of significance
   c. at the 0.01 level of significance

3. We draw a sample of size 50 from a population and observe a sample mean of 90 and a sample standard deviation of 4. Test the hypothesis that the population mean = 91.2

   a. at the 0.10 level of significance
   b. at the 0.05 level of significance
   c. at the 0.01 level of significance

4. We draw a sample of size 35 from a population and observe a sample mean of 25.5 and a sample standard deviation of 2.2. Test the hypothesis that the population mean = 26

   a. at the 0.10 level of significance
   b. at the 0.05 level of significance
   c. at the 0.01 level of significance

5. We draw a sample of size 90 from a population and observe a sample mean of 100.5 and a sample standard deviation of 2.1. Test the hypothesis that the population mean = 100

   a. at the 0.10 level of significance
   b. at the 0.05 level of significance
   c. at the 0.01 level of significance

6. We draw a sample of size 1000 from a population and observe a sample mean of 100.1 and a sample standard deviation of 2.1. Test the hypothesis that the population mean = 100

   a. at the 0.10 level of significance
   b. at the 0.05 level of significance
   c. at the 0.01 level of significance

7. We draw a sample of size 2000 from a population and observe a sample mean of 100.1 and a sample standard deviation of 2.1. Test the hypothesis that the population mean = 100

   a. at the 0.10 level of significance
   b. at the 0.05 level of significance
   c. at the 0.01 level of significance

8. We draw a sample of size 3000 from a population and observe a sample mean of 100.1 and a sample standard deviation of 2.1. Test the hypothesis that the population mean = 100

   a. at the 0.10 level of significance
   b. at the 0.05 level of significance
   c. at the 0.01 level of significance

PROBLEM SET C

1. Does observing 22 defectives out of a sample of 300 cause you to reject the claim that the true proportion of defectives is 0.05 or less

   a. at the 0.10 level of significance?
   b. at the 0.05 level of significance?
   c. at the 0.01 level of significance?

2. Does observing 20 defectives out of a sample of 300 cause you to reject the claim that the true proportion of defectives is 0.05 or less

   a. at the 0.10 level of significance?
   b. at the 0.05 level of significance?
   c. at the 0.01 level of significance?

3. Does observing 30 defectives out of a sample of 300 cause you to reject the claim that the true proportion of defectives is 0.08 or less

   a. at the 0.10 level of significance?
   b. at the 0.05 level of significance?
   c. at the 0.01 level of significance?

4. Does observing 12 defectives out of a sample of 100 cause you to reject the claim that the true proportion of defectives is 0.10 or less

   a. at the 0.10 level of significance?
   b. at the 0.05 level of significance?
   c. at the 0.01 level of significance?

5. Does observing 15 defectives out of a sample of 100 cause you to reject the claim that the true proportion of defectives is 0.10 or less

   a. at the 0.10 level of significance?
   b. at the 0.05 level of significance?
   c. at the 0.01 level of significance?

6. Does observing 16 defectives out of a sample of 100 cause you to reject the claim that the true proportion of defectives is 0.10 or less

   a. at the 0.10 level of significance?
   b. at the 0.05 level of significance?
   c. at the 0.01 level of significance?

7. Does observing 17 defectives out of a sample of 100 cause you to reject the claim that the true proportion of defectives is 0.10 or less

   a. at the 0.10 level of significance?
   b. at the 0.05 level of significance?
   c. at the 0.01 level of significance?

8. Does observing 18 defectives out of a sample of 100 cause you to reject the claim that the true proportion of defectives is 0.10 or less

   a. at the 0.10 level of significance?

b. at the 0.05 level of significance?

c. at the 0.01 level of significance?

9. Does observing 50 defectives out of a sample of 1000 cause you to reject the claim that the true proportion of defectives is 0.04 or less

a. at the 0.10 level of significance?

b. at the 0.05 level of significance?

c. at the 0.01 level of significance?

10. Does observing 50 defectives out of a sample of 1000 cause you to reject the claim that the true proportion of defectives is 0.03 or less

a. at the 0.10 level of significance?

b. at the 0.05 level of significance?

c. at the 0.01 level of significance?

PROBLEM SET D

1. A sample of size 150 is drawn from population A, and a sample mean of 72.3 and a sample standard deviation of 2.50 are observed. Then a sample of size 170 is drawn from population B, and a sample mean of 71.8 and sample standard deviation of 2.64 are observed. Test the hypothesis that the mean of population A is greater than the mean of population B.

a. at the 0.10 level of significance

b. at the 0.05 level of significance

c. at the 0.01 level of significance

2. A sample of size 50 is drawn from population A, and a sample mean of 42 and a sample standard deviation of 2.80 are observed. Then a sample of size 65 is drawn from population B, and a sample mean of 41.3 and sample standard deviation of 3.20 are observed. Test the hypothesis that the mean of population A is greater than the mean of population B.

a. at the 0.10 level of significance

b. at the 0.05 level of significance

c. at the 0.01 level of significance

3. A sample of size 35 is drawn from population A, and a sample mean of 88.1 and a sample standard deviation of 6.30 are observed. Then a sample of size 115 is drawn from population B, and a sample mean of 85.8 and sample standard deviation of 5.20 are observed. Test the hypothesis that the mean of population A is greater than the mean of population B.

a. at the 0.10 level of significance

b. at the 0.05 level of significance

c. at the 0.01 level of significance

4. A sample of size 500 is drawn from population A, and a sample mean of 1.07 and a sample standard deviation of 0.50 are observed. Then a sample of size 600

is drawn from population B, and a sample mean of 1.03 and sample standard deviation of 0.10 are observed. Test the hypothesis that the mean of population A is greater than the mean of population B.

a. at the 0.10 level of significance
b. at the 0.05 level of significance
c. at the 0.01 level of significance

5. A sample of size 30 is drawn from population A, and a sample mean of 153.1 and a sample standard deviation of 2.60 are observed. Then a sample of size 223 is drawn from population B, and a sample mean of 152.8 and sample standard deviation of 2.80 are observed. Test the hypothesis that the mean of population A is greater than the mean of population B.

a. at the 0.10 level of significance
b. at the 0.05 level of significance
c. at the 0.01 level of significance

6. A sample of size 150 is drawn from population A, and a sample mean of 1020 and a sample standard deviation of 45.4 are observed. Then a sample of size 200 is drawn from population B, and a sample mean of 1005.2 and sample standard deviation of 34.2 are observed. Test the hypothesis that the mean of population A is greater than the mean of population B.

a. at the 0.10 level of significance
b. at the 0.05 level of significance
c. at the 0.01 level of significance

7. A sample of size 85 is drawn from population A, and a sample mean of 32.7 and a sample standard deviation of 29.9 are observed. Then a sample of size 115 is drawn from population B, and a sample mean of 30.7 and sample standard deviation of 50.3 are observed. Test the hypothesis that the mean of population A is greater than the mean of population B.

a. at the 0.10 level of significance
b. at the 0.05 level of significance
c. at the 0.01 level of significance

8. A sample of size 1700 is drawn from population A, and a sample mean of 101.1 and a sample standard deviation of 3.50 are observed. Then a sample of size 2000 is drawn from population B, and a sample mean of 100.9 and sample standard deviation of 2.60 are observed. Test the hypothesis that the mean of population A is greater than the mean of population B.

a. at the 0.10 level of significance
b. at the 0.05 level of significance
c. at the 0.01 level of significance

9. A sample of size 35 is drawn from population A, and a sample mean of 21.7 and a sample standard deviation of 0.13 are observed. Then a sample of size 30 is

drawn from population B, and a sample mean of 21.6 and sample standard deviation of 0.12 are observed. Test the hypothesis that the mean of population A is greater than the mean of population B.

a. at the 0.10 level of significance
b. at the 0.05 level of significance
c. at the 0.01 level of significance

PROBLEM SET E

1. A sample of size 150 is drawn from population A, and a sample mean of 72.3 and a sample standard deviation of 2.50 are observed. Then a sample of size 170 is drawn from population B, and a sample mean of 71.8 and sample standard deviation of 2.64 are observed. Test the hypothesis that there is no difference between the population means

   a. at the 0.10 level of significance
   b. at the 0.05 level of significance
   c. at the 0.01 level of significance

2. A sample of size 50 is drawn from population A, and a sample mean of 42 and a sample standard deviation of 2.80 are observed. Then a sample of size 65 is drawn from population B, and a sample mean of 41.3 and sample standard deviation of 3.20 are observed. Test the hypothesis that there is no difference between the population means

   a. at the 0.10 level of significance
   b. at the 0.05 level of significance
   c. at the 0.01 level of significance

3. A sample of size 35 is drawn from population A, and a sample mean of 88.1 and a sample standard deviation of 6.30 are observed. Then a sample of size 115 is drawn from population B, and a sample mean of 85.8 and sample standard deviation of 5.20 are observed. Test the hypothesis that there is no difference between the population means

   a. at the 0.10 level of significance
   b. at the 0.05 level of significance
   c. at the 0.01 level of significance

4. A sample of size 500 is drawn from population A, and a sample mean of 1.07 and a sample standard deviation of 0.50 are observed. Then a sample of size 600 is drawn from population B, and a sample mean of 1.03 and sample standard deviation of 0.10 are observed. Test the hypothesis that there is no difference between the population means

   a. at the 0.10 level of significance
   b. at the 0.05 level of significance
   c. at the 0.01 level of significance

5. A sample of size 30 is drawn from population A, and a sample mean of 153.1 and a sample standard deviation of 2.60 are observed. Then a sample of size 223 is drawn from population B, and a sample mean of 152.8 and sample standard deviation of 2.80 are observed. Test the hypothesis that there is no difference between the population means

   a. at the 0.10 level of significance
   b. at the 0.05 level of significance
   c. at the 0.01 level of significance

6. A sample of size 150 is drawn from population A, and a sample mean of 1005.2 and a sample standard deviation of 45.4 are observed. Then a sample of size 200 is drawn from population B, and a sample mean of 1020 and sample standard deviation of 34.2 are observed. Test the hypothesis that there is no difference between the population means

   a. at the 0.10 level of significance
   b. at the 0.05 level of significance
   c. at the 0.01 level of significance

7. A sample of size 85 is drawn from population A, and a sample mean of 30.7 and a sample standard deviation of 29.9 are observed. Then a sample of size 115 is drawn from population B, and a sample mean of 32.7 and sample standard deviation of 50.3 are observed. Test the hypothesis that there is no difference between the population means

   a. at the 0.10 level of significance
   b. at the 0.05 level of significance
   c. at the 0.01 level of significance

8. A sample of size 1700 is drawn from population A, and a sample mean of 101.1 and a sample standard deviation of 3.50 are observed. Then a sample of size 2000 is drawn from population B, and a sample mean of 100.9 and sample standard deviation of 2.60 are observed. Test the hypothesis that there is no difference between the population means

   a. at the 0.10 level of significance
   b. at the 0.05 level of significance
   c. at the 0.01 level of significance

9. A sample of size 35 is drawn from population A, and a sample mean of 21.7 and a sample standard deviation of 0.13 are observed. Then a sample of size 30 is drawn from population B, and a sample mean of 21.6 and sample standard deviation of 0.12 are observed. Test the hypothesis that there is no difference between the population means

   a. at the 0.10 level of significance
   b. at the 0.05 level of significance
   c. at the 0.01 level of significance

PROBLEM SET F   1. For the following contingency table, compute

a. chi-square
b. the number of degrees of freedom
c. the significance level

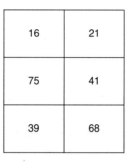

16	21
75	41
39	68

2. For the following contingency table, compute

a. chi-square
b. the number of degrees of freedom
c. the significance level

15	15	40
35	39	47

3. For the following contingency table, compute

a. chi-square
b. the number of degrees of freedom
c. the significance level

307	311	155
45	55	35

4. For the contingency table at the top of the next page, compute

a. chi-square
b. the number of degrees of freedom
c. the significance level

40	51	32
22	33	54
13	15	22

5. For the following contingency table, compute
   a. chi-square
   b. the number of degrees of freedom
   c. the significance level

51	106	107
40	40	53
39	56	42

6. For the following contingency table, compute
   a. chi-square
   b. the number of degrees of freedom
   c. the significance level

12	56	52	108
15	55	51	98
16	63	57	99

7. For the contingency table at the top of the next page, compute
   a. chi-square
   b. the number of degrees of freedom
   c. the significance level

15	15	31	61
20	21	45	88
21	19	44	78
106	110	219	399

8. For the following contingency table, compute
   a. chi-square
   b. the number of degrees of freedom
   c. the significance level

15	15	31	61
20	21	45	88
21	19	44	78
399	219	110	106

CHAPTER 4 QUIZ   1. The hypothesis that we can test directly is called the _____ .

2. The "other" hypothesis—the one that states that the null hypothesis is false—is
   called the _____ .

3. When we want to compute the probability of 69 or more successes out of 100
   trials, $p = 0.60$, we compute a z-score

$$z = \frac{68.5 - 60}{\sqrt{(100)(0.60)(0.40)}} .$$

   That we compute with 68.5 rather than 69 is called the _____

   _____ of the normal approximation to the binomial distribution.

4. Rejecting the null hypothesis when in fact it is true is called "making a _____ _____ ."

5. If I want to test the hypothesis $\mu \leq 10$, I am making a _____ test.

6. If I want to test the hypothesis $\mu = 10$, I am making a _____ test.

7. In the cells of a contingency table, the numbers we enter are always integers (whole numbers) because they represent _____ _____ .

8. In using chi-square for a test on a contingency table, the null hypothesis is *always* that _____ _____ .

9. The 5% critical value of a statistic is the value where 5% of the area of the distribution _____ _____ .

ANSWERS TO EXERCISES

1. "Fair coin" means $p = \frac{1}{2}$.

a. *"More than 55"* means 56, 57, 58, 59, 60, . . . , 100. The histogram looks like:

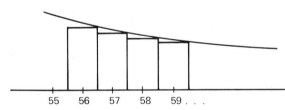

55   56   57   58   59 . . .

So the z-score is to be computed, after the continuity correction, on $55\frac{1}{2}$ (because that's where the bar for 56 starts).

$$n = 100; \quad p = \frac{1}{2}; \quad np = 50; \quad z = \frac{55\frac{1}{2} - 50}{\sqrt{(100)(\frac{1}{2})(\frac{1}{2})}} = 1.1$$

From Table 1, Prob($z \geq 1.1$) = 0.1357:

$$\text{Prob}(z \geq 1.1) = \text{Probability that } z \text{ is greater than } 1.1$$
$$= 0.5000 - \text{Prob}(0 \leq z \leq 1.1),$$

which, from Table 1, is

$$0.5000 - 0.3643 = 0.1357.$$

b. For 40 or more heads, we compute from 39½:

$$z = \frac{39\frac{1}{2} - 50}{\sqrt{(100)(\frac{1}{2})(\frac{1}{2})}} = -2.1$$

$$\text{Prob}(z \geqslant -2.1) = 0.50 + 0.4821 = 0.9821$$

2. Same as Exercise 1, but $p = 0.4$; $np = 40$.

a.
$$z = \frac{55\frac{1}{2} - 40}{\sqrt{100(0.4)(0.6)}} = 3.16$$

Table 1 gives values for z-scores only to $z = 3.09$.
$\text{Prob}(z \geqslant 3.09) = 0.5000 - 0.4990 = 0.001$, so $\text{Prob}(z \geqslant 3.16) < 0.001$.

b.
$$z = \frac{39\frac{1}{2} - 40}{\sqrt{100(0.4)(0.6)}} = -0.10$$

$$\text{Prob}(z \geqslant -0.10) = 0.50 + 0.0398 = 0.5398$$

3. To test $p \leqslant 0.05$, test $p = 0.05$ one-tailed. We observe 30 defectives, $n = 300$.
If $p = 0.05$, 30 converts to the z-score of

$$z = \frac{(30 - \frac{1}{2}) - np}{\sqrt{np(1 - p)}} = \frac{29.5 - (0.05)(300)}{\sqrt{300(0.05)(0.95)}} = 3.84.$$

(Note the continuity correction.) Thus,

$$\text{Prob}(30 \text{ } or \text{ } more \text{ defectives}) = \text{Prob}(z \geqslant 3.84) < 0.001.$$

So reject $p = 0.05$ at the 0.001 level of significance. Since any $p < 0.05$ leads to an even higher z-score

$$[\text{for example, if } p = 0.04. \text{ then: } \quad z = \frac{29.5 - (0.04)(300)}{\sqrt{300(0.04)(0.96)}} = 5.16],$$

we reject all $p \leqslant 0.05$, at the 0.001 level of significance.

4. I: Buy insurance and don't have an accident (Cost: $500; Prob = 0.98).

   II: Don't buy insurance and have an accident (Cost: $10,000; Prob = 0.02).

5. a. One-tailed; the claim is 5 minutes *or less*. If $\mu = 5.0$, the value of $\bar{x} = 5.8$ converts to the z-score

$$z = \frac{5.8 - 5.0}{3/\sqrt{40}} = 1.69.$$

Then

$$\text{Prob}(z \geqslant 1.69) = 0.5000 - 0.4545$$
$$= 0.0455 < 0.05.$$

Then reject 5.0 as a mean at the 0.05 level. Also, reject any $\mu < 5.0$ at the 0.05 level, as well (as the z-scores will only be *larger*).

b.

$$z = \frac{5.7 - 5.0}{3/\sqrt{40}} = 1.48$$

We can't reject the company's claim at the 0.05 level, since the *one-tailed* 0.05 critical value for z is 1.645. But we can reject at the 0.10 level, since $\text{Prob}(z \geqslant 1.48) = 0.5000 - 0.4306 = 0.0694 < 0.10$. (Usually, however, statisticians require that you set up the test *and* the level of significance before you make the computation.)

6.

$$z = \frac{36.3 - 37}{2.8/\sqrt{50}} = -1.77$$

We *cannot* reject (two-tailed). (See Figure 4-3.)

7. By the wording of the questions, part a is a one-tailed test ("better" versus "worse"), while part b is a two-tailed test ("no different" versus "different").

a. $z = \dfrac{5.0 - 4.8}{\sqrt{\dfrac{(5.3)^2}{40} + \dfrac{(4.6)^2}{45}}} = 0.18;$ not significant, one-tailed.

b. $z = 0.18;$ also not significant, two-tailed.

8. Compute row and column totals, then compute expected values:

	Republican and Others	Democrat	Total
Low	17 (28.08)	35 (23.92)	52
Middle	43 (37.80)	27 (32.20)	70
High	48 (42.12)	30 (35.88)	78
Total	108	92	200

$$\frac{108 \times 52}{200} = 28.08 \qquad \frac{92 \times 52}{200} = 23.92$$

$$\frac{108 \times 70}{200} = 37.80 \qquad \frac{92 \times 70}{200} = 32.20$$

$$\frac{108 \times 78}{200} = 42.12 \qquad \frac{92 \times 78}{200} = 35.88$$

$$\chi^2 = \frac{(17 - 28.08)^2}{28.08} + \frac{(35 - 23.92)^2}{23.92} + \frac{(43 - 37.80)^2}{37.80}$$

$$+ \frac{(27 - 32.20)^2}{32.20} + \frac{(48 - 42.12)^2}{42.12} + \frac{(30 - 35.88)^2}{35.88}$$

$$= 12.844$$

9. The number of degrees of freedom is

$$\text{(Number of rows} - 1) \times \text{(Number of columns} - 1) = (3 - 1) \times (2 - 1)$$
$$= 2 \times 1 = 2 \text{ d.f.}$$

The critical values for χ^2, 2 d.f., are 5.991, 7.378, 9.210, and 10.597, from Table 4. The value of chi-square we observe is 12.844, which exceeds the ½% critical value $\chi^2_{0.005}$; therefore, 12.844 is in the 0.5% tail. So we reject the null hypothesis—that the row and column variables are not related—at the 0.005 level. [We would write: "The value of $\chi^2 = 12.844$ is significant ($P < 0.005$)."]

ANSWERS TO CHAPTER 4 QUIZ

1. null hypothesis

2. alternate hypothesis

3. continuity correction

4. type 1 error

5. one-tailed

6. two-tailed

7. the counts of individuals who fall in that category

8. there is no relationship between the row and column variables

9. lies to the right

Small-Sample Inference: The t-Distribution

Terms we'll be learning about in this chapter

t-distribution

t-test

5-1 ESTIMATING THE MEAN

In the inferences we've made in the previous chapters, knowing the distribution of the statistic allowed us to form confidence intervals for the parameter we were estimating. In the case of confidence intervals for the mean, we knew that intervals

$$\bar{x} \pm 1.96\sigma_{DSM} \tag{1}$$

would capture the mean of the population 95% of the time. The problem was, we couldn't know σ_{DSM} without knowing the population standard deviation σ_{pop}, since

$$\sigma_{DSM} = \frac{\sigma_{pop}}{\sqrt{n}} \tag{2}$$

So what we did was take s, the sample standard deviation, as the value of the population standard deviation σ_{pop} in (2) and compute 95% confidence intervals

$$\bar{x} \pm 1.96\frac{s}{\sqrt{n}}. \tag{3}$$

We said that as long as n, the sample size, was 30 or more, simply substituting the value of s for σ_{pop} would not affect the confidence intervals in any significant way. We also said that if the sample size were 30 or more, the DSM would be approximately normal for virtually all populations we would study.

But when the sample size is less than 30, the DSM will no longer be approximately normal, unless the population is approximately normal to begin with. So in this chapter, we will make the assumption that the population from which we are drawing samples is approximately normal. (Don't be too concerned about this. Frequently, the kinds of populations from which our numbers come will be approximately normal populations. And statisticians have spent a lot of effort studying how good all these approximations are if the population fails to be approximately normal—an area of study called **robust statistics**. Don't worry about these details. Others have worried about them for you.)

Now, it's a fact that when n is 30 or larger, the substitution of s for σ_{pop} is good enough that

$$\bar{x} \pm 1.96\frac{s}{\sqrt{n}} \tag{4}$$

captures the population mean 95% of the time. But when n is less than 30, there is some uncertainty in replacing σ_{pop} by s. So we would expect that to capture the population mean 95% of the time, we would have to form an interval that is *wider* than $1.96(s/\sqrt{n})$. And the smaller the sample size n, the more uncertainty, so the wider the interval. This is, in fact, the case. With sample size 10, for example, a

95% confidence interval for the mean is $\bar{x} \pm 2.26(s/\sqrt{n})$. With a sample size of 4, a 95% confidence interval is $\bar{x} \pm 3.18(s/\sqrt{n})$. A complete table of so-called *t*-values is given in Table 3 at the back of the book.

5-2 HOW TO USE THE *t*-TABLE

The *t*-table is constructed so as to replace the 1.96σ figure you're accustomed to by appropriately *larger* values when the sample size is less than 30. When you turn to the table, you will notice that along the left-hand side of the table is a column marked *n* for sample size, and adjacent to it is a column marked d.f., which stands for degrees of freedom. Degrees of freedom is a technical term, invented by statisticians, that has to do with how many "freely varying" items there are as we make our estimates. Don't concern yourself further with the term d.f., except to note that it's the same d.f. you encountered in chi-square, but in the case of the *t*-distribution for one sample or one population the degrees of freedom is equal to 1 less than the sample size, or $n - 1$. You'll encounter the concept of d.f. one more time, when you study the *F*-distribution in Chapter 6.

The *t*-table is actually a table of critical values of the *t*-distribution, just as the chi-square table is a table of critical values of the chi-square distribution. Reprinted below is the third line of the *t*-table:

n	d.f.	$t_{0.100}$	$t_{0.050}$	$t_{0.025}$	$t_{0.010}$	$t_{0.005}$
4	3	1.638	2.353	3.182	4.541	5.841

As Figure 5-1a on page 180 indicates (for sample size 4), 10% (= 0.10) of the area of the *t*-distribution lies to the right of the value 1.638. Therefore, 1.638 is the 10%, or 0.10, critical value for *t*, 3 d.f., labeled $t_{0.10}$. Figure 5-1b records that 5% of the same *t*-distribution lies to the right of 2.353; that is, $t_{0.05} = 2.353$. As Figure 5-1 suggests, the *t*-distribution is symmetric about 0. While 10% of the distribution lies to the right of $+1.638$, 10% also lies to the *left* of -1.638. Similarly, 5% of the distribution lies to the left of -2.353.

The larger the number of degrees of freedom, the closer the *t*-distribution comes to the normal distribution. At 200 d.f., not shown in Table 3, the *t*-distribution looks just like the normal:

n	d.f.	$t_{0.100}$	$t_{0.050}$	$t_{0.025}$	$t_{0.010}$	$t_{0.005}$
201	200	1.282	1.645	1.960	2.326	2.576

This line is actually included in our *t*-table, as a line marked ∞. At this point, the *t*-distribution equals the normal distribution exactly. Notice that for d.f. 200, the 5%

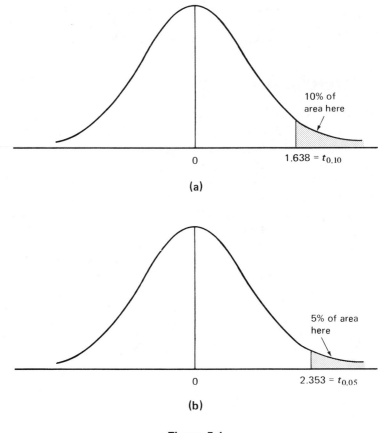

Figure 5-1

critical value, $t_{0.05}$, is exactly what we find for the 5% critical value for the normal distribution, 1.645. That is, we know that the probability of finding a z-score to the right of (greater than) 1.645 is 0.05, and similarly, the probability of finding a z-score larger than 1.960 is 0.025. Since the normal distribution is symmetric about 0, the probability of a z-score less than -1.96 is also 0.025. Adding, we get $0.025 + 0.025 = 0.05 = 5\%$. That's why the value 1.96 has come to be known as the **5%, two-tailed critical value for the normal distribution**.

EXERCISE 1

For each of the following sets of data, calculate the sample mean $\bar{x}$ and the sample variance s^2. Then, using the t-table (Table 3), form 90%, 95%, and 99% confidence intervals for the population mean. (Remember that you need to get two tails of the distribution.) Assume that the populations from which the samples are drawn are approximately normal in each case.

a. 3, 3, 4, 4, 5, 6

b. 20, 20, 25, 25, 26, 27, 38, 41

c. 100, 101, 101, 102, 102, 105, 106, 108, 109, 110, 110, 111

d. 1, 1, 1, 2, 2, 3, 3, 3, 3, 4, 4, 4, 5, 5, 6, 6, 7, 7, 7, 8, 8, 8, 9, 9, 10, 10, 10

e. 5, 5, 5, 6, 6, 6, 6, 6, 7, 7, 7, 7, 7, 8, 8, 8, 8, 9, 9, 9, 10, 10, 10, 10, 11, 11, 11, 12, 12, 12, 13, 13, 14, 15, 15, 16, 18

5-3 HYPOTHESIS TESTING

We know that for a normal population, when we draw a small sample (less than 30) and estimate the population variance by the sample variance, we must adjust, by means of the t-distribution, for the extra uncertainty in our estimate. We can apply the same ideas to hypothesis testing. Suppose we have a sample—10, 10, 11, 12—from what is assumed (or known) to be a normal population. We can test the hypothesis $\mu = 13$ against the alternative $\mu \neq 13$ at the 0.05 level of significance. We call such tests **t-tests**. We compute a t-score, just like a z-score:

$$t = \frac{\bar{x} - 13}{s/\sqrt{n}} = \frac{10.75 - 13}{0.957/\sqrt{4}} = -4.70.$$

Since $|t| = 4.70$ is greater than 3.182, the 0.025 critical value for t, $4 - 1 = 3$ d.f., we reject the null hypothesis $\mu = 13$ and conclude that $\mu \neq 13$.

We choose the critical value $t_{0.025}$ to make our 0.05 level of significance test, because we are making a two-tailed test. If we were making a one-tailed test, $\mu \geq 13$ versus $\mu < 13$, we would use the $t_{0.05}$ critical value, 2.353, instead.

Example 1

Use the same small-sample data—10, 10, 11, 12—assuming it comes from a normal population. Test the hypothesis $\mu < 12$ at the 0.05 and 0.01 levels of significance.

Solution

We test $\mu < 12$ by considering the alternative $\mu \geq 12$, setting up as the null hypothesis $\mu = 12$. We make a one-tailed test, and if we can reject $\mu = 12$, we can reject $\mu > 12$, as well, and conclude that $\mu < 12$. See Section 4-3, page 144, for the details of the reasoning. Computing the t-score,

$$t = \frac{10.75 - 12}{0.957/\sqrt{4}} = -2.61.$$

From the table, $t_{0.05} = 2.353$ and $t_{0.01} = 4.541$. Ignoring the minus sign in our t-value, we see that t exceeds the 0.05 level critical value, but does not exceed the 0.01 level critical value. We conclude that $\mu < 12$ at the 0.05 level of significance. ■

EXERCISE 2

Suppose you have a sample of 5 measurements: 10, 10, 10, 11, 12.

a. Assuming a normal population, can you conclude that $\mu < 12$ (one-tailed)?

b. Can you conclude that $\mu \neq 12$ (two-tailed)?

5-4 TESTING DIFFERENCES BETWEEN MEANS

Suppose we have two populations, A and B, and we draw samples A and B from the two populations. On the basis of these two samples, can we conclude that the means of the two populations are different? We must assume that the two populations A and B are both normal, and that both have the same (but unknown) variance. To test the hypothesis $\mu_A = \mu_B$ against either the two-sided alternative $\mu_A \neq \mu_B$ or either of the one-sided alternatives $\mu_A < \mu_B$ or $\mu_B < \mu_A$, we proceed exactly as we did in Section 4-5, except that we compute a t-score instead of a z-score. The t-statistic we compute involves a complicated expression in the denominator because of the weighted estimates for the variance that come from the two samples:

$$t = \frac{\bar{x}_A - \bar{x}_B}{\sqrt{\left[\dfrac{\Sigma(x_A - \bar{x}_A)^2 + \Sigma(x_B - \bar{x}_B)^2}{n_A + n_B - 2}\right]\left(\dfrac{1}{n_A} + \dfrac{1}{n_B}\right)}}$$

and t has $(n_A + n_B - 2)$ degrees of freedom. If $n_A + n_B - 2 > 30$, use the last line in Table 3 (at the end of the book), marked ∞. The expression $\Sigma(x_A - \bar{x}_A)^2$ stands for the sum of the squared distances (deviations) of the elements of sample A from the mean of sample A. Similarly for sample B. Also, n_A is the sample size for sample A, and n_B is the sample size for sample B.

Example 2

An educator is studying the difference in verbal abilities between boys and girls. On a test, the boys scored 10, 10, 11. The girls scored 10, 12, 12, 13. Do the data indicate there is a difference in mean verbal abilities between the boys and the girls as measured by the test at the 5% level? Assume that both populations are normal and have the same variance.

Solution

$$
\text{Boys:} \quad \bar{x} = 10.33; \quad \Sigma(x - \bar{x})^2 = 2s^2 = 0.67
$$
$$
\text{Girls:} \quad \bar{x} = 11.75; \quad \Sigma(x - \bar{x})^2 = 3s^2 = 4.75
$$

$$
t = \frac{10.33 - 11.75}{\sqrt{\left[\dfrac{0.67 + 4.75}{3 + 4 - 2}\right]\left(\dfrac{1}{3} + \dfrac{1}{4}\right)}} = \frac{-1.42}{0.795} = -1.786
$$

The two-tailed 5% critical value at $(3 + 4 - 2) = 5$ d.f. is $t_{0.025} = 2.571$. The value -1.786 is not less than -2.571, so the difference is not significant. ∎

EXERCISE 3   In Example 2, what if the boys scored 10, 10, 10, 11 and the girls scored 10, 12, 12, 12, 12? Make the same assumptions as in Example 2.

PROBLEM SET A

1. Assume the following sample of size 5 comes from a normal population. Find 90%, 95%, and 99% confidence intervals for the mean.

    49   59   60   56   62

2. Assume the following sample of size 6 comes from a normal population. Find 90%, 95%, and 99% confidence intervals for the mean.

    42.92   41.59   44.56   56.39   60.61   64.48

3. Assume the following sample of size 4 comes from a normal population. Find 90%, 95%, and 99% confidence intervals for the mean.

    72   70   71   71

4. Assume the following sample of size 15 comes from a normal population. Find 90%, 95%, and 99% confidence intervals for the mean.

    1.7   14.0   11.4   14.6   6.0   8.2   14.4   11.9   8.3   11.7
    5.5   6.4   14.1   8.9   12.2

5. Assume the following sample of size 29 comes from a normal population. Find 90%, 95%, and 99% confidence intervals for the mean.

    36.7   33.2   19.5   38.5   33.5   34.1   24.1   33.9   25.5   29.6
    29.4   24.0   18.8   15.6   30.3   25.3   34.0   18.8   29.8   30.4
    38.6   26.6   23.7   29.0   33.9   23.4   16.1   34.5   25.1

6. Assume the following sample of size 12 comes from a normal population. Find 90%, 95%, and 99% confidence intervals for the mean.

    137.51   36.69   36.09   57.07   46.61   70.54   51.14   22.10
    160.93   59.31   89.56   92.66

7. Assume the following sample of size 18 comes from a normal population. Find 90%, 95%, and 99% confidence intervals for the mean.

17.1	46.0	50.6	50.9	44.5	24.6	52.4	34.6	33.2
39.7	23.2	33.7	49.7	14.8	37.4	27.0	28.4	4.2

8. Assume the following sample of size 22 comes from a normal population. Find 90%, 95%, and 99% confidence intervals for the mean.

181,000	66,000	112,000	139,000	139,000	160,000	159,000
142,000	108,000	178,000	188,000	208,000	214,000	75,000
80,000	215,000	111,000	132,000	167,000	120,000	123,000
64,000						

9. Assume the following sample of size 8 comes from a normal population. Find 90%, 95%, and 99% confidence intervals for the mean.

54	58	23	51	83	81	2	88

10. Assume the following sample of size 26 comes from a normal population. Find 90%, 95%, and 99% confidence intervals for the mean.

287	346	326	290	315	300	274	326	340	325
265	235	297	326	299	268	330	331	285	314
381	315	302	323	298	259				

PROBLEM SET B

1. Assume the following sample of size 5 comes from a normal population. Test the hypothesis that the mean of the population is 65 or less

   a. at the 0.10 level of significance
   b. at the 0.05 level of significance
   c. at the 0.01 level of significance

67	71	68	69	65

2. Assume the following sample of size 6 comes from a normal population. Test the hypothesis that the mean of the population is 55 or less

   a. at the 0.10 level of significance
   b. at the 0.05 level of significance
   c. at the 0.01 level of significance

60.80	61.92	63.01	58.03	56.14	56.31

3. Assume the following sample of size 4 comes from a normal population. Test the hypothesis that the mean of the population is 70 or less

   a. at the 0.10 level of significance
   b. at the 0.05 level of significance
   c. at the 0.01 level of significance

73	71	71	72

4. Assume the following sample of size 15 comes from a normal population. Test the hypothesis that the mean of the population is 9 or less

   a. at the 0.10 level of significance
   b. at the 0.05 level of significance
   c. at the 0.01 level of significance

12.8	5.8	14.4	10.2	5.2	12.8	14.1	12.4	7.7	10.5
7.4	5.1	13.7	12.2	5.7					

5. Assume the following sample of size 29 comes from a normal population. Test the hypothesis that the mean of the population is 25 or less

   a. at the 0.10 level of significance
   b. at the 0.05 level of significance
   c. at the 0.01 level of significance

34.6	16.8	36.4	33.8	34.6	20.9	19.1	32.6
28.7	22.5	18.6	27.5	19.9	26.9	26.8	19.4
26.0	20.2	18.6	35.2	36.0	24.1	36.5	24.0
23.4	32.4	31.8	33.5	39.9			

6. Assume the following sample of size 12 comes from a normal population. Test the hypothesis that the mean of the population is 75 or less

   a. at the 0.10 level of significance
   b. at the 0.05 level of significance
   c. at the 0.01 level of significance

108.40	159.01	44.13	27.95	148.98	38.41	122.55
42.90	33.55	156.47	173.08	157.13		

7. Assume the following sample of size 18 comes from a normal population. Test the hypothesis that the mean of the population is 15 or less

   a. at the 0.10 level of significance
   b. at the 0.05 level of significance
   c. at the 0.01 level of significance

11.0	49.4	32.2	25.0	47.6	26.9	10.7	52.2	38.4
9.1	0.8	19.8	16.9	18.7	43.8	6.1	43.3	50.6

8. Assume the following sample of size 22 comes from a normal population. Test the hypothesis that the mean of the population is 100,000 or less

   a. at the 0.10 level of significance
   b. at the 0.05 level of significance
   c. at the 0.01 level of significance

221,000	212,000	144,000	85,000	183,000	200,000
57,000	218,000	213,000	128,000	86,000	193,000
142,000	107,000	179,000	46,000	51,000	218,000
196,000	71,000	153,000	152,000		

9. Assume the following sample of size 8 comes from a normal population. Test the hypothesis that the mean of the population is 20 or less

   a. at the 0.10 level of significance
   b. at the 0.05 level of significance
   c. at the 0.01 level of significance

21	40	65	58	26	3	87	9

10. Assume the following sample of size 26 comes from a normal population. Test the hypothesis that the mean of the population is 285 or less

    a. at the 0.10 level of significance
    b. at the 0.05 level of significance
    c. at the 0.01 level of significance

237	267	333	309	315	259	262	317	283	274
286	315	338	292	334	343	288	270	299	323
278	327	349	313	265	323				

PROBLEM SET C

1. Assume the following sample of size 5 comes from a normal population. Test the hypothesis that the mean of the population is 70

   a. at the 0.10 level of significance
   b. at the 0.05 level of significance
   c. at the 0.01 level of significance

68	67	67	71	64

2. Assume the following sample of size 6 comes from a normal population. Test the hypothesis that the mean of the population is 65

   a. at the 0.10 level of significance
   b. at the 0.05 level of significance
   c. at the 0.01 level of significance

56.83	65.81	59.32	60.18	57.56	58.96

3. Assume the following sample of size 4 comes from a normal population. Test the hypothesis that the mean of the population is 70

   a. at the 0.10 level of significance
   b. at the 0.05 level of significance
   c. at the 0.01 level of significance

72	71	71	72

4. Assume the following sample of size 15 comes from a normal population. Test the hypothesis that the mean of the population is 11

   a. at the 0.10 level of significance
   b. at the 0.05 level of significance
   c. at the 0.01 level of significance

10.8	10.7	13.0	14.5	11.7	14.8	13.9	5.6	10.2	7.0
6.3	13.6	12.9	11.2	9.6					

5. Assume the following sample of size 29 comes from a normal population. Test the hypothesis that the mean of the population is 30

   a. at the 0.10 level of significance
   b. at the 0.05 level of significance
   c. at the 0.01 level of significance

24.1	15.9	36.6	16.5	36.7	38.4	20.2	18.8
26.7	33.5	20.3	18.1	36.2	39.4	30.4	35.4
22.0	24.8	27.0	18.3	37.2	19.4	29.1	35.9
19.3	24.8	39.4	31.4	34.1			

6. Assume the following sample of size 12 comes from a normal population. Test the hypothesis that the mean of the population is 125

   a. at the 0.10 level of significance
   b. at the 0.05 level of significance
   c. at the 0.01 level of significance

131.43	32.42	116.40	93.64	64.40	114.67	106.26
94.26	31.97	23.71	23.86	167.28		

7. Assume the following sample of size 18 comes from a normal population. Test the hypothesis that the mean of the population is 15

   a. at the 0.10 level of significance
   b. at the 0.05 level of significance
   c. at the 0.01 level of significance

11.7	15.1	20.1	35.6	39.7	31.4	1.3	14.8	28.9
31.9	2.7	14.9	29.1	42.7	2.5	38.0	11.1	8.8

8. Assume the following sample of size 22 comes from a normal population. Test the hypothesis that the mean of the population is 100,000

   a. at the 0.10 level of significance
   b. at the 0.05 level of significance
   c. at the 0.01 level of significance

75,000	201,000	76,000	44,000	140,000	108,000
98,000	73,000	166,000	93,000	182,000	111,000
130,000	189,000	149,000	205,000	82,000	218,000
98,000	120,000	199,000	151,000		

9. Assume the following sample of size 8 comes from a normal population. Test the hypothesis that the mean of the population is 70

   a. at the 0.10 level of significance
   b. at the 0.05 level of significance
   c. at the 0.01 level of significance

85	20	65	60	17	44	16	65

10. Assume the following sample of size 26 comes from a normal population. Test the hypothesis that the mean of the population is 285
   a. at the 0.10 level of significance
   b. at the 0.05 level of significance
   c. at the 0.01 level of significance

289	357	289	277	267	328	296	339	355	291
266	288	303	348	327	329	279	311	282	318
348	294	377	317	206	259				

PROBLEM SET D

1. The following sample of size 3 is drawn from population *A*, which is assumed to be approximately normal:

   73.88   71.49   73.15

   The following sample of size 4 is drawn from population *B*, which is also assumed to be approximately normal:

   72.48   69.99   70.06   71.62

   Assume that the two populations have the same variance. Test the hypothesis that the two populations have the same mean
   a. at the 0.10 level of significance
   b. at the 0.05 level of significance
   c. at the 0.01 level of significance

2. The following sample of size 5 is drawn from population *A*, which is assumed to be approximately normal:

   41.6   41.9   42.2   42.1   42.9

   The following sample of size 6 is drawn from population *B*, which is also assumed to be approximately normal:

   40.0   40.0   41.6   40.8   41.5   41.9

   Assume that the two populations have the same variance. Test the hypothesis that the two populations have the same mean
   a. at the 0.10 level of significance
   b. at the 0.05 level of significance
   c. at the 0.01 level of significance

3. The following sample of size 4 is drawn from population *A*, which is assumed to be approximately normal:

   91.0   84.3   86.4   89.0

   The following sample of size 11 is drawn from population *B*, which is also assumed to be approximately normal:

   73.5   78.2   74.8   75.6   67.8   69.9   76.1   71.0   73.9
   76.6   78.8

   Assume that the two populations have the same variance. Test the hypothesis that the two populations have the same mean

a. at the 0.10 level of significance
b. at the 0.05 level of significance
c. at the 0.01 level of significance

4. The following sample of size 3 is drawn from population A, which is assumed to be approximately normal:

$$1.37 \quad 1.44 \quad 1.12$$

The following sample of size 12 is drawn from population B, which is also assumed to be approximately normal:

1.19   1.18   1.25   1.29   1.23   1.22   1.16   1.20   1.13   1.22
1.11   1.25

Assume that the two populations have the same variance. Test the hypothesis that the two populations have the same mean

a. at the 0.10 level of significance
b. at the 0.05 level of significance
c. at the 0.01 level of significance

5. The following sample of size 12 is drawn from population A, which is assumed to be approximately normal:

156.7   155.2   152.8   151.2   154.0   151.6   151.1   153.3   154.3
151.4   155.0   152.5

The following sample of size 22 is drawn from population B, which is also assumed to be approximately normal:

148.6   151.2   150.3   150.6   149.9   150.4   149.0   149.9   152.6
152.9   150.1   149.0   153.1   151.1   151.8   154.5   153.0   151.2
149.9   149.4   150.0   148.0

Assume that the two populations have the same variance. Test the hypothesis that the two populations have the same mean

a. at the 0.10 level of significance
b. at the 0.05 level of significance
c. at the 0.01 level of significance

6. The following sample of size 15 is drawn from population A, which is assumed to be approximately normal:

1037.3   998.7   1058.6   973.0   995.2   965.4   1066.3
1014.4   997.8   969.4   996.1   998.4   977.5   1019.3
1034.2

The following sample of size 20 is drawn from population B, which is also assumed to be approximately normal:

1019.6   992.2   1063.3   1001.9   1039.9   1049.7   1005.9
1018.1   1013.4   998.6   987.1   1032.1   1045.9   1017.2
1043.9   1031.1   1014.0   1008.7   1030.0   1002.5

Assume that the two populations have the same variance. Test the hypothesis that the two populations have the same mean

a. at the 0.10 level of significance
b. at the 0.05 level of significance
c. at the 0.01 level of significance

7. The following sample of size 5 is drawn from population *A*, which is assumed to be approximately normal:

$$20.3 \quad 25.6 \quad 11.1 \quad 35.4 \quad 50.4$$

The following sample of size 15 is drawn from population *B*, which is also assumed to be approximately normal:

$$40.1 \quad 27.6 \quad 65.0 \quad 22.8 \quad 15.4 \quad 10.4 \quad -6.3 \quad 12.9 \quad 16.8 \quad 61.7$$
$$29.3 \quad 48.9 \quad 26.1 \quad 11.1 \quad 38.5$$

Assume that the two populations have the same variance. Test the hypothesis that the two populations have the same mean

a. at the 0.10 level of significance
b. at the 0.05 level of significance
c. at the 0.01 level of significance

CHAPTER 5 QUIZ

1. We use the *t*-distribution instead of the normal when the _____ is less than 30.

2. If the sample size is 30 or more, we can use the _____ rather than the _____ .

3. (True-False) For small-sample confidence intervals, we use the *t*-distribution, but for small-sample one- and two-tailed hypothesis testing, we must use the normal distribution.

4. Four uses of the *t*-table for small-sample statistics are

_____

_____

_____

_____

ANSWERS TO EXERCISES

1. a. $n = 6$, 5 d.f.; $\bar{x} = 4.167$; $s = 1.169$; $\dfrac{s}{\sqrt{n}} = 0.477$;

$t_{0.05} = 2.015$
90%: $\bar{x} \pm t_{0.05} \dfrac{s}{\sqrt{n}} = 4.167 \pm (2.015)(0.477)$
$= 4.167 \pm 0.962$
$= (3.205, 5.129)$

$t_{0.025} = 2.571$
95%: $4.167 \pm (2.571)(0.477) = (2.941, 5.393)$

$t_{0.005} = 4.032$
99%: $(2.24, 6.09)$

b. $n = 8$, 7 d.f.; $\bar{x} = 27.75$; $s = 7.74$; $\dfrac{s}{\sqrt{n}} = 2.74$;

90%: $27.75 \pm 5.19 = (22.56, 32.94)$
95%: $27.75 \pm 6.47 = (21.28, 34.22)$
99%: $27.75 \pm 9.58 = (18.17, 37.33)$

c. $n = 12$, 11 d.f.; $\bar{x} = 105.42$; $s = 4.1$; $\dfrac{s}{\sqrt{n}} = 1.18$;

90%: $105.42 \pm 2.13 = (103.29, 107.55)$
95%: $105.42 \pm 2.61 = (102.82, 108.02)$
99%: $105.42 \pm 3.68 = (101.74, 109.10)$

d. $n = 27$, 26 d.f.; $\bar{x} = 5.41$; $s = 2.96$; $\dfrac{s}{\sqrt{n}} = 0.57$;

90%: $5.41 \pm 0.97 = (4.44, 6.38)$
95%: $5.41 \pm 1.17 = (4.24, 6.58)$
99%: $5.41 \pm 1.59 = (3.82, 7.00)$

e. $n > 30$, so use normal theory or line marked ∞ in t-table: $n = 37$; $\bar{x} = 9.51$;

$s = 3.37$; $\dfrac{s}{\sqrt{n}} = 0.55$;

90%: $9.51 \pm 1.645(0.55) = 9.51 \pm 0.91 = (8.60, 10.42)$
95%: $9.51 \pm 1.96(0.55) = 9.51 \pm 1.09 = (8.42, 10.60)$
99%: $9.51 \pm 2.576(0.55) = 9.51 \pm 1.43 = (8.08, 10.94)$

2. $\bar{x} = 10.6$; $\dfrac{s}{\sqrt{n}} = 0.4$; $\dfrac{\bar{x} - 12}{s/\sqrt{n}} = \dfrac{10.6 - 12}{0.4} = -3.5$

a. *One-tailed*:
4 d.f., $t_{0.05} = 2.132$. Significant at 0.05 level.
4 d.f., $t_{0.01} = 3.747$. Not significant at 0.01 level.
So, $\mu < 12$ at the 0.05 level.

b. *Two-tailed*:
4 d.f., $t_{0.025} = 2.776$. Significant at 0.05 level.
4 d.f., $t_{0.005} = 4.604$. Not significant at 0.01 level.
So, $\mu \neq 12$ at the 0.05 level.

3. Boys: $\bar{x} = 10.25$; $\Sigma(x - \bar{x})^2 = 3s^2 = 0.75$
Girls: $\bar{x} = 11.6$; $\Sigma(x - \bar{x})^2 = 4s^2 = 3.2$

$$t = \frac{10.25 - 11.6}{\sqrt{\left(\dfrac{0.75 + 3.2}{4 + 5 - 2}\right)\left(\dfrac{1}{4} + \dfrac{1}{5}\right)}} = \frac{-1.35}{0.504} = -2.68$$

Two-tailed 5% critical value, at 7 d.f., $t_{0.025} = 2.365$. Since $-2.68 < -2.365$ (two-tailed), the difference is significant at the 0.05 level. Or, since 2.68 exceeds 2.365 (dropping the minus signs), the difference is significant at the 0.05 level.

ANSWERS TO
CHAPTER 5 QUIZ

1. sample size

2. normal distribution; t-distribution

3. False: we use the t-distribution for hypothesis testing as well.

4. confidence intervals; one-tailed hypothesis testing; two-tailed hypothesis testing; hypothesis testing—differences between means

Analysis of
Variance (ANOVA)

Terms we'll be learning about in this chapter

ANOVA table

blocks

F-statistic

matrix

treatments

variance ratio

6-1 DO ALL THE POPULATIONS HAVE THE SAME MEAN?

Analysis of variance, or ANOVA, was originally developed so that agricultural researchers could determine which kind of fertilizer had the best effect on crop yields. The procedure, of course, applies to many other situations, as well: we're going to be testing lifetimes of different brands of light bulbs shortly, but you can substitute tons per acre for hours and different kinds of fertilizer for brands of light bulb. ANOVA can be used, in certain situations, to test whether or not several populations all have the same mean.

Suppose we have three different brands of light bulb, brands A, B, and C. And suppose we wish to test whether or not there is any difference between the lifetimes of these three brands. We select a sample of four bulbs from each of the three brands and observe how long each bulb lasts. Suppose the results are as shown in Figure 6-1. Take a good look at these numbers. Most people without statistical knowledge would recognize that these data indicate that bulb C lasts longer than bulb B, which lasts longer than bulb A. In particular, the three different brands do *not* all have the same average lifetime.

Now suppose instead that the results of the test were those shown in Figure 6-2. In this case, we would probably conclude that the data do not show a significant

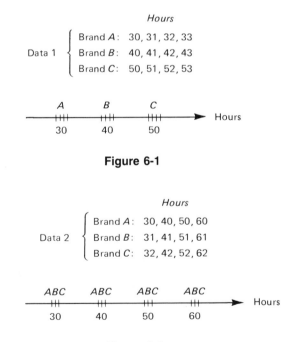

Figure 6-1

Figure 6-2

difference between the mean bulb lifetimes for brands A, B, and C. Just why is this? For data 2, we cannot conclude that the means are different because the amount of *variation within brands* is larger than the amount of *variation across brands*; for data 1, on the other hand, the variation across brands is much larger than the variation within brands. This is the basic idea of analysis of variance: to compare the across-brand variation with the within-brand variation.

We will always make two assumptions about the populations (A, B, and C) when we do ANOVA:

1. The populations are all normally distributed, or approximately normally distributed.

2. All the populations have the same variance σ^2, which in general will not be known or given to us.

Under these assumptions, the samples of bulb lifetimes that gave rise to data 1 and 2 might come from populations that look like the ones in Figure 6-3. It's pretty clear on intuitive grounds that for data 1 the population means are different, while for data 2 we can't necessarily conclude that.

The null hypothesis we will be testing is that these normal populations (in our example, three normal populations) all have the same mean. (If the null hypothesis

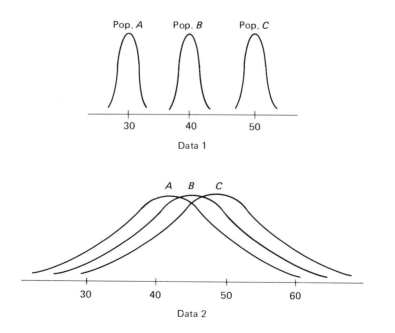

Figure 6-3

is true, then the three normal populations are all the *same* normal population, since we've already assumed that the populations all have the same variance.)

The method of testing this null hypothesis is to compute two estimates for the (common) variance of the three populations in two different ways. The first way involves the assumption (the null hypothesis) that the means are all the same. The second way doesn't. If the two estimates for the variance are too far apart, the assumption that the population means are all the same must be wrong, so we reject the null hypothesis. If the two estimates are not too far apart, we don't reject the null hypothesis. To decide what "too far apart" means, we need a statistic and its distribution, assuming the null hypothesis that the population means are all the same. That statistic is the **F-statistic**, or **F-ratio** as it is sometimes called. The *F*-ratio measures whether the two estimates are "far apart" or "close together" and is computed as

$$F = \frac{\text{Between-samples estimate}}{\text{Within-samples estimate}}.$$

First let's calculate *F* for data 1. For the within-samples estimate of the variance, take each of the three sample variances and pool (average) them:

$$\text{Mean square error (MSE)} = \frac{s_A^2 + s_B^2 + s_C^2}{3}$$

$$= \frac{1.67 + 1.67 + 1.67}{3}$$

$$= 1.67,$$

where s_A^2 stands for the variance of sample A, and so on.

For the between-samples estimate of the variance, consider the variance of the set of three sample means:

$$\text{var}(\overline{x}_A, \overline{x}_B, \overline{x}_C) = \text{var}(31.5, 41.5, 51.5).$$

Use your calculator to compute the variance of a *sample* of three numbers: {31.5, 41.5, 51.5}. Did you get 100? You should have.

$$\text{var}(31.5, 41.5, 51.5) = \frac{(31.5 - 41.5)^2 + (41.5 - 41.5)^2 + (51.5 - 41.5)^2}{3 - 1 = 2}$$

$$= 100.0.$$

The variance of the sample means for samples of size 4 is ¼ the variance of the population (see page 97), or, put another way, the population variance is 4 times the variance of the sample mean (for sample size 4). Thus $4 \times 100 = 400$ is an estimate of the population variance. Now we form the *ratio* of these two estimates of variance, to obtain our statistic F (and now you can see why it's called an F-ratio):

$$F = \frac{\left(\begin{array}{c} \text{Estimate of variance based on} \\ \text{variation between sample means} \end{array}\right)}{\left(\begin{array}{c} \text{Estimate of variance based on} \\ \text{variance with samples} \end{array}\right)} \qquad (1)$$

$$= \frac{\text{Between-samples variance estimate}}{\text{Within-samples variance estimate}}$$

$$= \frac{4(100)}{1.67} = 240.0.$$

Now both the numerator and denominator are estimates for the same thing—the population's common variance. So we should expect it to be about 1.0, on average. If we expect F to be 1 on average, is 240 too large to expect to see by chance? If we knew, say, a 5% critical value for F, we would know how to interpret 240. You may remember from Chapter 4, where we talked about chi-square (see pages 155–156), that the 5% critical value is the value on the number line such that 5% of the distribution is larger and 95% is smaller. It's like being in the 95th percentile of all values in the distribution (see Figure 6-4).

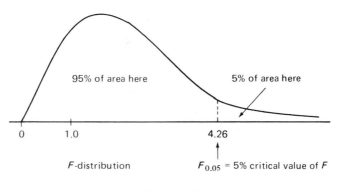

Figure 6-4

There is a different F-distribution for each *pair* of degrees of freedom (d.f.): (numerator d.f., denominator d.f.). The d.f. are based on N, the total number of observations (for example, $N = 12$ light bulbs tested), and k, the number of populations (for example, $k = 3$ brands of light bulb):

$$\text{Numerator d.f.} = k - 1,$$

$$\text{Denominator d.f.} = N - k.$$

In our example above,

$$\text{Numerator d.f.} = k - 1 = 3 - 1 = 2,$$

$$\text{Denominator d.f.} = N - k = 12 - 3 = 9.$$

We usually denote this (2, 9) d.f. If you look in Table 5, the table of critical values of F (at the back of the book), you will find that the 0.05 critical value at (2, 9) d.f. is $F_{0.05} = 4.26$. We already put the value 4.26 on Figure 6-4. We calculated an F-ratio of 240. If the null hypothesis is true, and all three populations have the same mean, then 5% of the time we will observe a value for the F-ratio as large as 4.26 or larger. We observed the value 240. We can reject the null hypothesis at the 0.05 level. Probably at the 0.00000001· level! This confirms our basic intuition about data 1, that the means are not all the same ($P < 0.05$).

Now let's calculate an F-ratio for data 2. For these data, $\bar{x}_A = 45$, $\bar{x}_B = 46$, $\bar{x}_C = 47$.

$$\text{var}(\bar{x}_A, \bar{x}_B, \bar{x}_C) = \text{var}(45, 46, 47) = \underline{\hspace{2cm}}$$

Take out your calculator and compute the variance of three sample means. Did you get a variance of 1.0? This is the between-samples (numerator) variance estimate. Then compute

$$s_A^2 = \text{var}(30, 40, 50, 60) = \underline{\hspace{3cm}},$$

$$s_B^2 = \text{var}(31, 41, 51, 61) = \underline{\hspace{3cm}},$$

$$s_C^2 = \text{var}(32, 42, 52, 62) = \underline{\hspace{3cm}}.$$

You should have gotten 166.67 for each of these three variances. They are all the same because the spread of the three samples is all the same. Each of the samples is shifted by 1.0 from the preceding one. Then we compute the F-ratio as follows:

$$F = \frac{\text{Between-samples variance}}{\text{Within-samples variance}}$$

$$= \frac{4\,\text{var}(\bar{x}_A, \bar{x}_B, \bar{x}_C)}{\dfrac{s_A^2 + s_B^2 + s_C^2}{3}} = \frac{4(1)}{166.67} = 0.024.$$

When we compute $F = 0.024$ with $F_{0.05} = 4.26$, we see that 0.024 is *not* significant at the 0.05 level. So we cannot conclude that the means are different, which confirms our intuition about data 2.

The critical values for F are found in Table 5. There are two parts of this table. Part a contains the 5% critical values; part b, the 1% critical values. Each set takes

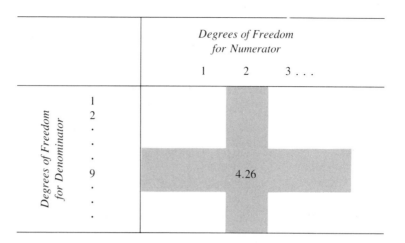

Figure 6-5   5% critical value for *F*: at (2, 9) d.f., $F_{0.05}$ = 4.26

up a whole page because *F* is indexed by two parameters, the numerator degrees of freedom and the denominator degrees of freedom. Figure 6-5 shows how part a of Table 5, the 5% critical values, is arranged. Take a moment and look up $F_{0.05}$ and $F_{0.01}$, (2, 9) d.f., in Table 5. You should see 4.26 and 8.02, respectively. If you don't, perhaps you are looking up (9, 2) degrees of freedom by mistake. The numerator d.f. (which in this case is 2) runs along the *top* of the table, while the denominator d.f. (in this case, 9) runs along the left-hand side of the table.

6-2 ANOVA TABLES AND UNEQUAL SAMPLE SIZES

We've just seen one method for computing the *F*-statistic. In practice, however, the computation is performed another way, via what is called an **ANOVA table** (see Figure 6-6, page 200).

One advantage of this new method of computing the *F*-ratio is that it permits ANOVA with *unequal sample sizes* as well. Without this method, if we had unequal sample sizes, we would have to "throw away" some of the data in some of the samples, until our samples were all of the same size; this would be wasteful and inefficient. The more data we have, the better inference we can make.

Returning to our ANOVA table (Figure 6-6), you can see that the table is divided into two rows, Treatment and Error, which end up being the numerator and denominator of *F*. Treatment originally referred to the effect of giving different fertilizers to different plots of land. In our light bulb example, it refers to the different brands of light bulb. In general, **treatment** refers to the between-samples vari-

Source of Variation	d.f.	Sum of Squares	Mean Square	F-Ratio
Treatment	$k - 1$	SS(Tr)	$MS(Tr) = \dfrac{SS(Tr)}{k - 1}$	$\dfrac{MS(Tr)}{MSE}$
	( )	( )	( )	( )
Error	$N - k$	SSE	$MSE = \dfrac{SSE}{N - k}$	
	( )	( )	( )	
Total	$N - 1$	SST		
	( )	( )		

Figure 6-6   ANOVA table

ance—the difference, if any, in population means. This will become the numerator of F. **Error** refers to the within-samples variance, which will become the denominator of F.

We begin computing d.f.:

$$k = \text{Number of populations;} \tag{2}$$

$$N = \text{Total number of data}$$

$$= \text{Sample size 1} + \text{Sample size 2} + \cdots + \text{Sample size } k. \tag{3}$$

For light bulb data 1, $k = 3$ populations (brands, in this case) and

$$N = 4 + 4 + 4 = 12.$$

Note also that sample size 1 = sample size 2 = sample size 3 = 4. We've been calling the samples A, B, and C. Now we're switching to 1, 2, and 3. We'll let n_1 = sample size 1; n_2 = sample size 2; and n_3 = sample size 3 to allow for the possibility that the sample sizes are all different. In our example, of course, $n_1 = n_2 = n_3 = 4$. We can now rewrite equation (3) as

$$N = n_1 + n_2 + n_3 + \cdots + n_k. \tag{4}$$

Don't be put off by (4). It says the same thing that (3) does, only in a more concise form. Take your time to look at (3) and (4) until you see it. Make sure you understand what n_1, n_2, and n_3 stand for. (What would n_{10} stand for? What would n_i stand for? What does the i mean in n_i? If you can't answer these questions, here are the answers: n_{10} = size of sample from population 10; n_i = size of sample from population i; i is the index, telling us *which* population or sample we're talking about; i must be an integer between 1 and k.)

Source of Variation	d.f.	Sum of Squares	Mean Square	F-Ratio
Treatment	$k - 1$	SS(Tr)	$MS(Tr) = \dfrac{SS(Tr)}{k - 1}$	$\dfrac{MS(Tr)}{MSE}$
	(2)	( )	( )	( )
Error	$N - k$	SSE	$MSE = \dfrac{SSE}{N - k}$	
	(9)	( )	( )	
Total	$N - 1$	SST		
	(11)	( )		

Figure 6-7

You now have enough information to fill in the d.f. on the ANOVA table in Figure 6-6, for light bulb data 1. In this example, $k - 1 = 3 - 1 = 2$; $N - k = 12 - 3 = 9$. Your ANOVA table should now look like Figure 6-7. But we still have several blanks to fill in. Treatment Sum of Squares, denoted SS(Tr), is computed as follows:

$$SS(Tr) = \sum_{i=1}^{k} \frac{T_i^2}{n_i} - \frac{T^2}{N} . \tag{5}$$

To compute this, you'll need to know how to interpret the symbols:

$$\sum_{i=1}^{k} \frac{T_i^2}{n_i} = \frac{T_1^2}{n_1} + \frac{T_2^2}{n_2} + \frac{T_3^2}{n_3} + \cdots + \frac{T_k^2}{n_k} . \tag{6}$$

To compute (5) or (6), you'll have to know what $T_1, T_2, \ldots, T_k$ are.

$T_1 = $ Sum of all the elements in the sample from population 1   (7)

To write (7) in symbols, you need to name all the elements of sample 1. They will be

$$x_{11}, x_{12}, x_{13}, \ldots, x_{1n_1} \tag{8}$$

Sample from population 1

There are n_1 elements in the sample

Labels for sample 1

Don't be put off by (8). Once you get used to the format, you'll see how useful it really is. And you'll find you have new abilities to read and understand technical material.

For light bulb data 1, the elements of sample 1 are 30, 31, 32, 33. In this case, n_1 = size of sample 1 = 4, and so

$$x_{11} = 30, \quad x_{12} = 31, \quad x_{13} = 32, \quad x_{14} = 33,$$

where x_{11} stands for the first element of sample 1, x_{12} stands for the second element of sample 1, and so on. (The order doesn't matter. We can make $x_{11} = 31$ and $x_{12} = 30$.) As long as we're naming the elements of sample 1, we might as well name the elements of sample 2. Can you guess how they would be labeled? [*Hint*: See equation (8).]

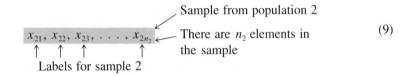

$$ \tag{9} $$

If you understand (8) as the labels of sample 1, and (9) as the labels of sample 2, can you guess what the labels would be for sample 3? In general, for sample i? For sample 3,

$$x_{31}, x_{32}, x_{33}, \ldots, x_{3n_3}. \tag{10}$$

For sample i,

$$x_{i1}, x_{i2}, x_{i3}, \ldots, x_{in_i}. \tag{11}$$

You may find it helpful to think of the elements of each of the k samples arranged in a rectangular array, each *row* of the array being the elements of one of the samples. [I've made the samples of different sizes in (12).]

$$
\begin{array}{llll llll}
x_{11} & x_{12} & x_{13} & \cdots & x_{1n_1} & \leftarrow & \text{Sample 1} \\
x_{21} & x_{22} & x_{23} & \cdots & x_{2n_2} & \leftarrow & \text{Sample 2} \\
x_{31} & x_{32} & x_{33} & \cdots & x_{3n_3} & \leftarrow & \text{Sample 3} \\
\vdots & \vdots & \vdots & & \vdots & & \\
& & & & & & \\
x_{i1} & x_{i2} & x_{i3} & \cdots & x_{in_i} & \leftarrow & \text{Sample } i \\
\vdots & \vdots & \vdots & & \vdots & & \\
& & & & & & \\
x_{k1} & x_{k2} & x_{k3} & \cdots & x_{kn_k} & \leftarrow & \text{Sample } k
\end{array}
\tag{12}
$$

In the case of data 1, this array would look like

$$
\begin{array}{llll}
30 & 31 & 32 & 33 & \leftarrow & \text{Sample 1} \\
40 & 41 & 42 & 43 & \leftarrow & \text{Sample 2} \\
50 & 51 & 52 & 53 & \leftarrow & \text{Sample 3}
\end{array}
\tag{13}
$$

Now look back at (7). Now that you know how to label the elements of sample 1, can you rewrite (7) without words? (Take a moment and write it in the space below.)

Rewritten: (7)

Let's see how what you wrote compares with what I would write:

$$
T_1 = \text{Sum of all the elements of sample 1}
$$
$$
= x_{11} + x_{12} + x_{13} + \cdots + x_{1n_1},
\tag{14}
$$

or, using the Σ sign,

$$
T_1 = \sum_{j=1}^{n_1} x_{1j}.
\tag{15}
$$

Now that we've defined T_1, can you guess that T_2 is defined as?

$$
T_2 = \text{Sum of all the elements of sample 2}
\tag{16}
$$

Use what you learned above in (14) and (15) to rewrite (16) in symbols:

Rewritten like (14): (16)

Rewritten like (15): (16)

Did you get

$$
T_2 = x_{21} + x_{22} + x_{23} + \cdots + x_{2n_2}
\tag{17}
$$

and

$$
T_2 = \sum_{j=1}^{n_2} x_{2j}?
\tag{18}
$$

(Hang in there. We're almost done with the symbols!)

Now that we have formulas for T_1 and T_2, can you write a formula for T_i, the total of the ith sample? Look at (14) and (15). Then look at (17) and (18). You should see what to change. Write your answer below:

$$T_i = \qquad\qquad\qquad\qquad\qquad\qquad\qquad\qquad\qquad (19)$$

Compare what you wrote above with what I wrote below:

$$T_i = x_{i1} + x_{i2} + x_{i3} + \cdots + x_{in_i} \qquad\qquad (20)$$

or

$$T_i = \sum_{j=1}^{n_i} x_{ij}. \qquad\qquad (21)$$

Notice that in (21) we have come to our first example of a two-variable subscript:

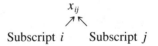

Subscript i Subscript j

Notice that in (20) and (21) the i is *fixed* (it tells us for which sample, namely, the ith sample, we are computing the total T_i), while the j *varies* (over all the elements of the ith sample to add them up).

Now we're almost ready to compute SS(Tr) using (5),

$$\text{SS(Tr)} = \sum_{i=1}^{k} \frac{T_i^2}{n_i} - \frac{T^2}{N}.$$

We know how to compute each T_i from (20) or (21). T is the *total* of all the elements, or the total of all the totals:

$$T = T_1 + T_2 + T_3 + \cdots + T_k \qquad\qquad (22)$$

or

$$T = \sum_{i=1}^{k} T_i. \qquad\qquad (23)$$

And, of course, n_i = the size of the sample from population i. Now, finally, we can compute SS(Tr). Use your calculator to calculate SS(Tr) for data 1. (The data are repeated at the top of the next page.) Fill in the blanks.

$$
\text{Data 1} \begin{cases} 30 & 31 & 32 & 33 & \leftarrow \text{Sample 1} \\ 40 & 41 & 42 & 43 & \leftarrow \text{Sample 2} \\ 50 & 51 & 52 & 53 & \leftarrow \text{Sample 3} \end{cases}
$$

$$
T_1 = \sum_{j=1}^{4} x_{1j} = x_{11} + x_{12} + x_{13} + x_{14} = \underline{\hspace{2cm}}
$$

$$
T_1^2 = \underline{\hspace{1.5cm}} \qquad \frac{T_1^2}{n_1} = \underline{\hspace{1.5cm}}
$$

$$
T_2 = \sum_{j=1}^{4} x_{2j} = x_{21} + x_{22} + x_{23} + x_{24} = \underline{\hspace{2cm}}
$$

$$
T_2^2 = \underline{\hspace{1.5cm}} \qquad \frac{T_2^2}{n_2} = \underline{\hspace{1.5cm}}
$$

$$
T_3 = \sum_{j=1}^{4} x_{3j} = x_{31} + x_{32} + x_{33} + x_{34} = \underline{\hspace{2cm}}
$$

$$
T_3^2 = \underline{\hspace{1.5cm}} \qquad \frac{T_3^2}{n_3} = \underline{\hspace{1.5cm}}
$$

$$
T = T_1 + T_2 + T_3 = \underline{\hspace{1.5cm}}
$$

$$
T^2 = \underline{\hspace{1.5cm}} \qquad N = \underline{\hspace{1.5cm}}
$$

$$
\frac{T^2}{N} = \underline{\hspace{1.5cm}} \qquad k = \underline{\hspace{1.5cm}}
$$

$$
\text{SS(Tr)} = \sum_{i=1}^{k} \frac{T_i^2}{n_i} - \frac{T^2}{N}
$$

$$
= \frac{T_1^2}{n_1} + \underline{\hspace{1.5cm}} + \underline{\hspace{1.5cm}} - \frac{T^2}{N}
$$

$$
= \underline{\hspace{1.5cm}}
$$

Did you get SS(Tr) = 800? If you didn't, you can check your computation with mine, shown as follows:

$$
T_1 = 30 + 31 + 32 + 33 = 126;
$$

$$
T_2 = 40 + 41 + 42 + 43 = 166;
$$

$$
T_3 = 50 + 51 + 52 + 53 = 206;
$$

$$T = 126 + 166 + 206 = 498; \quad N = 12; \quad k = 3;$$

$$\text{SS(Tr)} = \frac{T_1^2}{n_1} + \frac{T_2^2}{n_2} + \frac{T_3^2}{n_3} - \frac{T^2}{N}$$

$$= \frac{126^2}{4} + \frac{166^2}{4} + \frac{206^2}{4} - \frac{498^2}{12} = 800.$$

To compute SSE, Error Sum of Squares, we use the formula

$$\text{SSE} = \text{SST} - \text{SS(Tr)} \tag{24}$$

and while we already have SS(Tr), we need SST = Total Sum of Squares. SST is defined by

$$\text{SST} = \sum_{i,j} x_{ij}^2 - \frac{T^2}{N} . \tag{25}$$

This is the first time we've seen a summation sign with two indices (i, j) like

$$\sum_{i,j} x_{ij}^2. \tag{26}$$

This just means "add up the squares of *all* the x_{ij}." The x_{ij}'s are *all* the elements of *all* the samples. So in the case of our three samples of size 4, equation (26) just stands for

$$x_{11}^2 + x_{12}^2 + x_{13}^2 + x_{14}^2 + x_{21}^2 + x_{22}^2 + x_{23}^2 + x_{24}^2 + \cdots + x_{34}^2.$$

Now compute SST for data 1, using (25).

$$\text{SST} = \sum_{i,j} x_{ij}^2 - \frac{T^2}{N} = \underline{\hspace{2cm}}$$

Did you get 815? I calculated

$$\text{SST} = (30^2 + 31^2 + 32^2 + 33^2 + 40^2 + 41^2 + 42^2 + 43^2 + 50^2$$

$$+ 51^2 + 52^2 + 53^2) - \frac{498^2}{12}$$

$$= 21{,}482 - 20{,}667$$

$$= 815.$$

Then, from (24),

$$SSE = SST - SS(Tr)$$

$$= \underline{\quad} - \underline{\quad\quad}$$

$$= \underline{\quad} .$$

Did you get 15?

$$SSE = 815 - 800$$

$$= 15.$$

Now fill in the SS(Tr) and SSE entries in the ANOVA table. The table should now look like the one in Figure 6-8.

Source of Variation	d.f.	Sum of Squares	Mean Square	F-Ratio
Treatment	$k - 1$	SS(Tr)	$MS(Tr) = \dfrac{SS(Tr)}{k - 1}$	$\dfrac{MS(Tr)}{MSE}$
	(2)	(800)	( )	( )
Error	$N - k$	SSE	$MSE = \dfrac{SSE}{N - k}$	
	(9)	(15)	( )	
Total	$N - 1$	SST		
	(11)	(815)		

Figure 6-8

Completing the remainder of the table is easy. Just follow the formulas in the boxes; all the information you need is now in the table:

$$MS(Tr) = \frac{SS(Tr)}{k - 1} = \frac{800}{2} = 400,$$

$$MSE = \frac{SSE}{N - k} = \frac{15}{9} = 1.6667,$$

$$F = \frac{400}{1.6667} = 240.0.$$

Take this information and complete the table in Figure 6-8. The completed ANOVA table for data 1 should now look like the table in Figure 6-9.

Source of Variation	d.f.	Sum of Squares	Mean Square	F-Ratio
Treatment	$k - 1$	SS(Tr)	$MS(Tr) = \dfrac{SS(Tr)}{k-1}$	$\dfrac{MS(Tr)}{MSE}$
	(2)	(800)	(400)	(240)
Error	$N - k$	SSE	$MSE = \dfrac{SSE}{N-k}$	
	(9)	(15)	(1.6667)	
Total	$N - 1$	SST		
	(11)	(815)		

Figure 6-9

Let's review the computation we made and put all the formulas together:

$$T_i = \overset{n_i}{\underset{j=1}{\sum}} x_{ij} = x_{i1} + x_{i2} + \cdots + x_{in_i}, \qquad \overset{n_i \; \leftarrow \text{Size of } i\text{th sample}}{} \tag{27}$$

$$T = \overset{k}{\underset{i=1}{\sum}} T_i = T_1 + T_2 + \cdots + T_k, \qquad \overset{k \; \leftarrow \text{Number of populations}}{} \tag{28}$$

$$SST = \underset{i,j}{\sum} x_{ij}^2 - \frac{T^2}{N}, \tag{29}$$

$$SS(Tr) = \overset{k}{\underset{i=1}{\sum}} \frac{T_i^2}{n_i} - \frac{T^2}{N}, \tag{30}$$

$$SSE = SST - SS(Tr). \tag{31}$$

The remaining formulas are contained within the ANOVA table.

EXERCISE 1   Use the formulas on page 208 to compute the F-statistic for data 2 (page 194). Fill in the ANOVA table below:

Source of Variation	d.f.	Sum of Squares	Mean Square	F-Ratio
Treatment	$k - 1$ ( )	SS(Tr) ( )	$MS(Tr) = \dfrac{SS(Tr)}{k - 1}$ ( )	$\dfrac{MS(Tr)}{MSE}$ ( )
Error	$N - k$ ( )	SSE ( )	$MSE = \dfrac{SSE}{N - k}$ ( )	
Total	$N - 1$ ( )	SST ( )		

What is the null hypothesis for ANOVA in this case?

_____

Should we reject the null hypothesis for either data 1 or data 2 at either the 0.05 or 0.01 level of significance?

_____

_____

EXERCISE 2   What is the 0.05 critical value for F, (2, 9) d.f.? The 0.01 critical value?

_____   _____

What is the 0.05 critical value for F, (15, 6) d.f.? The 0.01 critical value?

_____   _____

EXERCISE 3   (Unequal sample sizes) I'd like to know whether or not my choice of bowling ball affects my score. My scores, the last 12 games, are

Ball A: 201   184   203

Ball B: 177   178   186   195

Ball C: 218   193   216   203   210

Does the ball affect my bowling score? Perform an analysis of variance on the three samples. Test the null hypothesis at the 0.05 level.

_____

_____

EXERCISE 4

To demonstrate that ANOVA works with more than three populations, suppose we add to data 2 (of page 194) a fourth brand of light bulb, brand *D*:

$$
\begin{array}{lllll}
\text{Brand } A: & 30 & 40 & 50 & 60 \\
\text{Brand } B: & 31 & 41 & 51 & 61 \\
\text{Brand } C: & 32 & 42 & 52 & 62 \\
\text{Brand } D: & 68 & 69 & 70 & 71 \\
\end{array}
$$

Test the hypothesis that the (population) means are all the same, at the 0.05 and 0.01 levels of significance.

6-3 TWO-WAY ANALYSIS OF VARIANCE (NO INTERACTION)

Recall our light bulb data of Figure 6-2 (page 194), data 2, which I've displayed again here:

Hours

$$
\text{Data 2}
\left\{
\begin{array}{llllll}
\text{Brand } A: & 30, & 40, & 50, & 60 \\
\text{Brand } B: & 31, & 41, & 51, & 61 \\
\text{Brand } C: & 32, & 42, & 52, & 62 \\
\end{array}
\right.
$$

Suppose that in this light bulb example, it happened that we were using light bulbs of different wattages as well as of different brands, say 60-watt light bulbs and 100-watt light bulbs. Let's look again at data 2 with this new information:

100-watt   *60-watt*

$$
\begin{array}{l}
\text{Data 2} \\
\text{(blocked)}
\end{array}
\left\{
\begin{array}{lllll}
\text{Brand } A: & 30, & 40 & 50, & 60 \\
\text{Brand } B: & 31, & 41 & 51, & 61 \\
\text{Brand } C: & 32, & 42 & 52, & 62 \\
\end{array}
\right.
$$

Before, we had three samples of size 4. Now we have six samples of size 2. The data are arranged in a **matrix**, with the **rows** representing the brands and the **columns** representing the different wattages, or sizes of bulb. The data are called **blocked** because the light bulbs are now not only arranged according to brand but also grouped into **blocks** within brand, according to wattage (60 or 100 watts). Notice that within each brand, the 60-watt bulbs tend to last longer than the 100-watt bulbs. Notice that the variation *within brand and wattage* is less than the variation within a given brand alone. By introducing a second factor, we have found a second suspected source of variation in bulb lifetimes. We are going to statistically verify or deny that either brand or wattage is in fact a source of variation in bulb lifetime. We are beginning a two-factor ANOVA or two-way analysis of variance.

The two factors (brands and wattage) are frequently referred to as **rows** and **columns**, or as **treatments** and **blocks**. We will also occasionally call them **factor A** and **factor B**. Notice now that each *cell*, or entry, in the matrix ("31, 41," for example) consists of a sample of two bulb lifetimes and can be identified by its row (brand, or treatment) and column (wattage, or block). The cell "31, 41" is (brand *B*, 100-watt), or the (2, 1)th entry, meaning its row is row 2 and its column is column 1:

		Column 1 ↓	
		100-watt	*60-watt*
	Brand *A*:	30, 40	50, 60
Row 2 →	Brand *B*:	31, 41	51, 61
	Brand *C*:	32, 42	52, 62

We have three rows and two columns, so we have six cells, or entries, in the matrix. Each entry represents a sample of two bulb lifetimes.

In order to perform a two-way analysis of variance, we are going to assume that the lifetime of the bulb depends in part on which brand of light bulb we have. We will call this the **treatment effect**. We will assume that a second part of the lifetime of the bulb depends on the wattage of the bulb. We will call this the **block effect**. The names "treatment effect" and "block effect" are what are commonly used in the statistical community, although it should be understood that the rows and columns could be interchanged equally well, so that the brand becomes the block effect and the wattage is the treatment. I would prefer using "factor *A*" and "factor *B*" for "treatment" and "block" because these names more adequately reflect the interchangeability of the two variables, but once again, I need to train you to converse in the statistical world. So, we'll call them "treatment" and "block" effects.

For the discussion that follows, we'll need to be familiar with several population parameters.

μ = Mean of all bulb lifetimes

μ_{ij} = Mean of bulb lifetimes from the (i, j)th cell, that is, the mean of the bulb lifetimes that are from the ith brand and the jth wattage. (For example, μ_{21} is the mean of brand *B*, 100-watt bulbs.)

a_i = Treatment effect of ith treatment

b_j = Block effect of jth block

If the treatment effects and block effects are simply additive to bulb lifetimes, which we shall assume, then we expect that

$$\mu_{ij} = \mu + a_i + b_j.$$

Now two-way ANOVA allows us to answer the two questions: (1) Do the different brands of light bulbs have the same or different mean lifetimes? and (2) Do different wattages of light bulbs have the same or different mean lifetimes?

If the brand has no effect on bulb lifetimes, then the population means for each brand of light bulb will be the same, and that will be the same as the overall population mean of all light bulbs. That is, the treatment effect of each treatment (brand) will be 0:

$$a_1 = a_2 = a_3 = 0.$$

If the wattage makes no difference, then the population means for each of the wattages will be the same, and again that will be the same as the overall population mean. That is, the block effect of each block (wattage) will be 0:

$$b_1 = b_2 = 0.$$

Whether the treatment effects and block effects are all 0 or not, we will *assume* that the mean of the population in the cell in row i, column j, the (i, j)th entry in the matrix, is given by the sum

$$\mu_{ij} = \mu + a_i + b_j. \tag{32}$$

This is called the assumption of **no interaction** between the treatment and block effects. We will have more to say about this later.

Now, the *two* null hypotheses of two-factor ANOVA are that the brand differences have no effect, and that the wattage differences have no effect, that is, that the treatment effects are all 0 and that the block effects are all 0.

The alternative to the first null hypothesis is that there is *some* difference in the mean lifetimes of the different brands of light bulb, and the alternative to the second null hypothesis is that there is some difference in the mean lifetimes of low- and high-wattage light bulbs. We will perform two F-tests, each very similar to the F-test we performed on our one-way ANOVA, and if the value of either of the statistics (F-ratios) is too large, we will reject the corresponding null hypothesis at the appropriate level of significance.

To construct the two-way ANOVA table, we'll need to introduce several new objects, and tell you how to compute them. None of this is any harder than it was in one-way ANOVA, but it's a little more complicated because we have more variables. As we said, we have a matrix of possible treatment–block combinations; in general, we will have r rows (treatments) and c columns (blocks). Let's assume that we have m observations in each cell of the matrix; that is, m observations for each treatment–block combination. We will study only the case of equal sample sizes

(m) for each cell in the treatment–block matrix, although two-way ANOVA can be performed on data of unequal sample sizes as well—the formulas just get more complicated. In our light bulb example, we have

$$r = 3 \text{ (Treatments, or rows),}$$

$$c = 2 \text{ (Blocks, or columns),}$$

$$m = 2 \text{ (Observations per cell).}$$

When we did one-way ANOVA, we needed two subscripts to describe our observations. Now that we have two factors, we'll need three subscripts. Don't panic! It's not especially difficult to deal with: Let x_{ijk} stand for the kth observation in the jth block and ith treatment group. So, $x_{211} = 31$ and $x_{212} = 41$:

Block 1
↓

	100-watt	60-watt
Brand *A*:	30, 40	50, 60

Observation 1
↓

Treatment 2 →	Brand *B*:	31, 41	51, 61

↑
Observation 2

	Brand *C*:	32, 42	52, 62

Now we need symbols for lots of different means: $\bar{x}_{ij.}$ stands for the mean of all the observations in the (i, j)th cell. The "." is placed in the subscript so you will know which of the subscripts, in this case the k, was allowed to vary in computing the mean. Another way to describe this computation is

$$\bar{x}_{ij.} = \frac{\sum_{k=1}^{m} x_{ijk}}{m} . \tag{33}$$

Now we can compute row means and column means, using the cell means:

$$\bar{x}_{i..} = \frac{\sum_{j=1}^{c} \bar{x}_{ij.}}{c} , \tag{34}$$

$$\bar{x}_{.j.} = \frac{\sum_{i=1}^{r} \bar{x}_{ij.}}{r} . \tag{35}$$

Next, we compute a "grand mean," which is the mean of all the data, and is also the mean of the row means or the mean of the column means:

$$\bar{x}_{...} = \frac{\sum_{i=1}^{r} \bar{x}_{i..}}{r} = \frac{\sum_{j=1}^{c} \bar{x}_{.j.}}{c}. \tag{36}$$

Now we can compute

$$SS(Tr) = cm \sum_{i=1}^{r} (\bar{x}_{i..} - \bar{x}_{...})^2, \tag{37}$$

$$SS(B) = rm \sum_{j=1}^{c} (\bar{x}_{.j.} - \bar{x}_{...})^2, \tag{38}$$

$$SSE = SST - SS(Tr) - SS(B) \tag{39}$$

(which we will explain shortly), and

$$SST = \sum_{i,j,k} (x_{ijk} - \bar{x}_{...})^2. \tag{40}$$

This new symbol,

$$\sum_{i,j,k}$$

just stands for "sum over all values of all three variables: i, j, and k." That is, the formula (40) for SST means "add up all the squared deviations

$$(x_{ijk} - \bar{x}_{...})^2$$

for all observations" (there are $r \cdot c \cdot m$ of them; in our light bulb example, $r \cdot c \cdot m = 3 \cdot 2 \cdot 2 = 12$).

The Sum of Squares for Error, SSE, is that part of the Total Sum of Squares which is not either SS(Tr) or SS(B). It assumes that there is "no interaction" between the factors (treatments and blocks), and this is what we shall be assuming. Actually, we made that assumption when we assumed that the a_i's and the b_j's were the treatment and block effects, respectively, and that the means of the cells were given by the formula (32). Had there been interaction between the factors, we could not have written the cell means by formula (32). By "interaction" we mean simply that one or more of the treatments behaves differently in the presence of one block condition than it does in the presence of a different block condition. Such behavior is commonly called **synergy**. As an example, suppose that one brand of light bulb was not a tungsten filament bulb, and that consequently the high-wattage bulbs lasted as long as the low-wattage bulbs for this brand. Then the assumption (32) would no longer be valid, and we would need to perform a two-way analysis of variance with interaction. The formulas for this become somewhat more complicated. For a further discussion, see Dunn and Clark, *Applied Statistics: Analysis of Variance and Regression* (New York: John Wiley & Sons, 1974).

We are in a position, finally, to compute the two-way ANOVA table (see Figure 6-10). Notice that it is very similar to the one-way ANOVA table.

Source of Variation	d.f.	Sum of Squares	Mean Square	F-Ratio	(d.f.)
Treatment	$r - 1$	SS(Tr)	$MS(Tr) = \dfrac{SS(Tr)}{r-1}$	$\dfrac{MS(Tr)}{MSE}$	$r - 1$ $rcm - m - c + 1$
Block	$c - 1$	SS(B)	$MS(B) = \dfrac{SS(B)}{c-1}$	$\dfrac{MS(B)}{MSE}$	$c - 1$ $rcm - m - c + 1$
Error	$rcm - m - c + 1$	SSE	$MSE = \dfrac{SSE}{rcm - m - c + 1}$		
Total	$rcm - 1$	SST			

Figure 6-10

In each cell in the rightmost column of the ANOVA table (Figure 6-10) is a pair of numbers written one over the other, which give the numerator and denominator degrees of freedom for the critical value of F. For the Treatment F-ratio, the degrees of freedom are the (treatment, error) d.f. For the Block F-ratio, the degrees of freedom are (block, error) d.f.

Now let's compute the ANOVA table for our light bulb example. The cell means $\bar{x}_{ij.}$ of the observations in the (i, j)th cell are as follows:

$$\bar{x}_{11.} = \frac{30 + 40}{2} = 35, \qquad \bar{x}_{12.} = \frac{50 + 60}{2} = 55,$$

$$\bar{x}_{21.} = \frac{31 + 41}{2} = 36, \qquad \bar{x}_{22.} = \frac{51 + 61}{2} = 56,$$

$$\bar{x}_{31.} = \frac{32 + 42}{2} = 37, \qquad \bar{x}_{32.} = \frac{52 + 62}{2} = 57.$$

The row means $\bar{x}_{i..}$ are the mean of the cell means for all the cells in the ith row:

$$\bar{x}_{1..} = \frac{35 + 55}{2} = 45, \qquad \bar{x}_{2..} = \frac{36 + 56}{2} = 46, \qquad \bar{x}_{3..} = \frac{37 + 57}{2} = 47.$$

The column means $\bar{x}_{.j.}$ are the means of the cell means for all the cells in the jth column:

$$\bar{x}_{.1.} = \frac{35 + 36 + 37}{3} = 36, \qquad \bar{x}_{.2.} = \frac{55 + 56 + 57}{3} = 56.$$

The grand mean $\bar{x}_{...}$ is the mean of the row means, and it is also the mean of the column means:

$$\bar{x}_{...} = \frac{36 + 56}{2} = \frac{45 + 46 + 47}{3} = 46.$$

Now, following (37),

$$\begin{aligned}
\text{SS(Tr)} &= c \cdot m \cdot [(\bar{x}_{1..} - \bar{x}_{...})^2 + (\bar{x}_{2..} - \bar{x}_{...})^2 + (\bar{x}_{3..} - \bar{x}_{...})^2] \\
&= 2 \cdot 2 \cdot [(45 - 46)^2 + (46 - 46)^2 + (47 - 46)^2] \\
&= 2 \cdot 2 \cdot [1 + 0 + 1] = 8.
\end{aligned}$$

Similarly, from (38),

$$\begin{aligned}
\text{SS(B)} &= r \cdot m \cdot [(\bar{x}_{.1.} - \bar{x}_{...})^2 + (\bar{x}_{.2.} - \bar{x}_{...})^2] \\
&= 3 \cdot 2 \cdot [(36 - 46)^2 + (56 - 46)^2] \\
&= 6 \cdot 200 = 1200.
\end{aligned}$$

From (40),

$$\begin{aligned}
\text{SST} =\ & (30 - 46)^2 + (40 - 46)^2 + (50 - 46)^2 + (60 - 46)^2 \\
& + (31 - 46)^2 + (41 - 46)^2 + (51 - 46)^2 + (61 - 46)^2 \\
& + (32 - 46)^2 + (42 - 46)^2 + (52 - 46)^2 + (62 - 46)^2 \\
=\ & 1508.
\end{aligned}$$

From (39),

$$\text{SSE} = 1508 - 8 - 1200 = 300.$$

Now we are ready to fill in our ANOVA table (Figure 6-11).

Source of Variation	d.f.	Sum of Squares	Mean Square	F-Ratio	(d.f.)
Treatment	2	8	4	0.10667	(2, 8)
Block	1	1200	1200	32	(1, 8)
Error	8	300	37.5		
Total	11	1508			

Figure 6-11

Since the tabular 0.05 critical value for F with (2, 8) d.f. is 4.46, while we observe a treatment value of $F = 0.10667$, we cannot reject the null hypothesis for the treatment—that all the $a_i = 0$—and so we conclude that the brand does not have a significant effect on bulb lifetime. (This is the same conclusion that we reached earlier, in our one-way ANOVA example for these same data, in Section 6-2.)

However, for the block, we find an F-ratio of 32, while the tabled 0.05 critical value for F with (1, 8) d.f. is 5.32. We conclude that the block effect (the wattage) is significant in the bulb lifetimes. In fact, the F-ratio exceeds the 0.01 critical value of 11.3.

These two results confirm our intuitive perception of the data in that we see considerable variation across wattage, when compared to within-cell variation, and we do not see significant variation across brand, when compared to within-cell variation.

EXERCISE 5

Take the light bulb data of Exercise 4, page 210, and perform a two-way ANOVA. Assume that there are four blocks (say four wattages: 150, 100, 60, and 40 watts), so the samples for each treatment–block cell in the matrix are of size 1:

	Watts			
	150	*100*	*60*	*40*
Brand *A*:	30	40	50	60
Brand *B*:	31	41	51	61
Brand *C*:	32	42	52	62
Brand *D*:	68	69	70	71

PROBLEM SET A   1. For the following data, compute an ANOVA table.

Sample of size 14 from population *A*:

14.2	12.7	11.7	11.2	13	13.2	11.1	12.4
10.9	11.4	12	12.8	12.6	13.9		

Sample of size 26 from population *B*:

15	14.5	14.4	12.9	16.3	14.9	15.3	15.3
12.8	14.4	15	14.3	15.3	13.5	16.7	17.5
15.8	15	16.1	12.9	13.7	15.5	13.1	13.4
13.5	13.6						

Then test the hypothesis that the populations all have the same mean at the 0.05 and 0.01 levels of significance.

2. For the following data, compute an ANOVA table.

Sample of size 3 from population *A*:

14.3   14.5   12.6

Sample of size 4 from population *B*:

13.2   10.7   12.4   13.1

Sample of size 6 from population *C*:

14.6   17.9   14.8   15   18.3   16

Then test the hypothesis that the populations all have the same mean at the 0.05 and 0.01 levels of significance.

3. For the following data, compute an ANOVA table.

Sample of size 8 from population *A*:

24.9   23.2   24.4   25.1   22.3   22.4   23.9   23

Sample of size 15 from population *B*:

25.8	24.1	25	24.1	25.6	25.5	24.9	24.6
23.3	25.8	26.3	24.7	25.1	24.4	24.2	

Sample of size 12 from population *C*:

23.1	24.5	20.8	21.2	22.2	22.3	21.7	23.5
21.7	23.5	19.9	23				

Sample of size 4 from population *D*:

24.6   24.4   25.1   22.6

Then test the hypothesis that the populations all have the same mean at the 0.05 and 0.01 levels of significance.

4. For the following data, compute an ANOVA table.

Sample of size 4 from population *A*:

51.7   53.2   52   52.7

Sample of size 9 from population B:

54.9   54.6   55.1   54.9   53.9   54.9   54.5   54.7
54.6

Sample of size 17 from population C:

56.2   55.6   54   55.2   56.2   53.2   53.6   53.4
53.4   54.1   54   52.4   53.3   55.9   54.8   51.8
53.3

Sample of size 23 from population D:

55   52   53   52.1   54.1   52.5   54.2   52.7
51.7   53.3   53.8   53.1   54.2   55.1   53.6   54.1
53.3   53.8   53.6   53.5   53.2   52.6   53.7

Sample of size 14 from population E:

54.7   53.5   53.9   54.6   53.7   53.2   54.4   53.6
53.3   53.3   52.8   54.5   53.5   55.9

Then test the hypothesis that the populations all have the same mean at the 0.05 and 0.01 levels of significance.

5. For the following data, compute an ANOVA table.

Sample of size 3 from population A:

17.2   18.4   16.6

Sample of size 4 from population B:

16.3   15.1   15.9   15.9

Sample of size 4 from population C:

16.1   16.2   16   15.9

Sample of size 5 from population D:

18.8   17.9   18   18   18.9

Sample of size 4 from population E:

16.1   17.4   18   18

Sample of size 3 from population F:

18.4   17.7   17.5

Sample of size 5 from population G:

17   16.5   18.4   16.3   16.3

Then test the hypothesis that the populations all have the same mean at the 0.05 and 0.01 levels of significance.

6. For the following data, compute an ANOVA table.

Sample of size 6 from population A:

45.1   45.1   46.8   44.5   46.5   42.8

Sample of size 15 from population B:

45	44.6	43.7	49.5	45	44.9	45.4	45.7
46.1	44.9	43	44	42.3	46.3	45.7	

Sample of size 12 from population C:

45.1	45	45.6	44.6	44.2	44.5	45.1	44.1
45.5	44.9	46.3	45.1				

Sample of size 10 from population D:

41.2	44.4	42.7	42.1	47	46.5	41.8	45.4
42.2	42.8						

Then test the hypothesis that the populations all have the same mean at the 0.05 and 0.01 levels of significance.

7. For the following data, compute an ANOVA table.

Sample of size 40 from population A:

53	57.4	53.3	54.9	55.5	53.7	53.3	55.1
55.7	52.3	55.5	55.2	55	53.5	53.4	53.7
58.2	55.1	56.7	54.5	54.1	54.1	56.7	56
55.9	56.4	54.3	54	52.8	54.1	56.8	54.4
53.6	56.3	54.5	56.6	55.2	54.8	54.7	55.8

Sample of size 30 from population B:

52.9	52.3	51.3	52.7	52.3	50.8	49.6	53
52.6	53.4	55.3	52.6	53.2	53.1	51.9	53.5
54.2	53.6	53.1	56	52	55.9	52.6	52.1
50.8	51.1	54	54.5	51	54.2		

Then test the hypothesis that the populations all have the same mean at the 0.05 and 0.01 levels of significance.

8. For the following data, compute an ANOVA table.

Sample of size 3 from population A:

13.72	14.63	15.95

Sample of size 4 from population B:

19.58	20.17	24.04	18.27

Then test the hypothesis that the populations all have the same mean at the 0.05 and 0.01 levels of significance.

9. For the following data, compute an ANOVA table.

Sample of size 3 from population A:

120	127	131

Sample of size 8 from population B:

196	178	213	196	190	172	186	202

Sample of size 6 from population C:

158   153   151   140   141   147

Then test the hypothesis that the populations all have the same mean at the 0.05 and 0.01 levels of significance.

CHAPTER 6 QUIZ

1. Analysis of variance was originally invented to make statistical tests on _____

_____ .

2. Analysis of variance is based on two estimates of _____ .

3. To use analysis of variance, all the populations must be _____ ,

or approximately _____ .

4. The F-ratio consists of dividing the _____

estimate of variance by the _____

estimate.

5. The null hypothesis of the F-statistic is that the populations all _____

_____ .

6. The value of F such that 5% of the distribution is greater than that value is called

the _____ .

7. There is a pair of degrees of freedom associated with the F-distribution, the

_____ degrees of freedom and the _____

degrees of freedom.

8. (True-False)   To use ANOVA, you must have equal sample sizes.

ANSWERS TO EXERCISES

1. Sample 1: $x_{11} = 30$, $x_{12} = 40$, $x_{13} = 50$, $x_{14} = 60$.
   Sample 2: $x_{21} = 31$, $x_{22} = 41$, $x_{23} = 51$, $x_{24} = 61$.
   Sample 3: $x_{31} = 32$, $x_{32} = 42$, $x_{33} = 52$, $x_{34} = 62$.

$$k = \text{Number of samples} = 3; \quad N = \text{Total number of data} = 12;$$

$$\bar{x}_1 = \frac{30 + 40 + 50 + 60}{4} = 45; \quad \bar{x}_2 = \frac{31 + 41 + 51 + 61}{4} = 46;$$

$$\bar{x}_3 = \frac{32 + 42 + 52 + 62}{4} = 47;$$

$$T_1 = 4(45) = 180; \quad T_2 = 4(46) = 184; \quad T_3 = 4(47) = 188;$$

$$T = 180 + 184 + 188 = 552;$$

$$SST = (30^2 + 40^2 + 50^2 + 60^2 + 31^2 + 41^2 + 51^2 + 61^2 + 32^2 + 42^2 + 52^2 + 62^2)$$

$$- \frac{552^2}{12}$$

$$= 26{,}900 - 25{,}392 = 1508;$$

$$SS(Tr) = \frac{180^2}{4} + \frac{184^2}{4} + \frac{188^2}{4} - \frac{552^2}{12}$$

$$= 25{,}400 - 25{,}392 = 8;$$

$$SSE = SST - SS(Tr) = 1508 - 8 = 1500;$$

$$\frac{SS(Tr)}{k-1} = \frac{8}{2} = 4; \quad \frac{SSE}{N-k} = \frac{1500}{9} = 166.7;$$

$$F = \frac{4}{166.7} = 0.024$$

The null hypothesis is that the populations all have the same mean. Since

$$F_{\text{data 2}} = 0.024 < F_{0.05} < F_{0.01} < 240 = F_{\text{data 1}},$$

reject the null hypothesis for data 1 at the 0.01 level. We cannot reject for data 2 at either level.

2. At (2, 9) d.f., $F_{0.05} = 4.26$; at (2, 9) d.f., $F_{0.01} = 8.02$;
at (15, 6) d.f., $F_{0.05} = 3.94$; at (15, 6) d.f., $F_{0.01} = 7.56$.
(If you didn't get these values, see if you *reversed* the numerator and denominator. The numerator is the first of the pair and goes along the top of the table, while the denominator is second and goes along the side of the table.)

3.

Source of Variation	d.f.	Sum of Squares	Mean Square	F-Ratio
Treatment	2	1284	642	6.83
Error	9	846	94	

Since $F_{0.05} = 4.26$ with (2, 9) d.f., 6.83 is significant at the 0.05 level. Reject the null hypothesis.

4. $F = 4.42$; at (3, 12) d.f., $F_{0.05} = 3.49$ and $F_{0.01} = 5.95$.
So, since 4.42 exceeds $F_{0.05}$, the F-ratio is significant at the 0.05 level. Since 4.42 does not exceed $F_{0.01}$, the F-ratio is not significant at the 0.01 level. So reject the null hypothesis (that means are all equal), $P < 0.05$.
(Look at the values in the exercise. You might think that it's "obvious" that the means are different, once we added the fourth population plus sample: 68, 69, 70, 71. The reason

the data fail to be significant at the 0.01 level is that the pooled estimate for the *common* variance σ^2 is large by virtue of the within-samples variance of samples A, B, and C.)

5.

Source of Variation	d.f.	Sum of Squares	Mean Square	F-Ratio	(d.f.)	Tabled Critical Value 0.05	0.01
Treatment	3	1664.75	554.9	16.44	(3, 9)	3.86	6.99
Block	3	1201.25	400.4	11.86	(3, 9)	3.86	6.99
Error	9	303.75	33.75				
Total	15	3169.75					

So we can conclude that both the treatments and the blocks are significant, at the 0.01 level of significance. That is, both brand and wattage are significant factors in the bulb lifetimes. Remember that when we treated these data in a one-way ANOVA, we found that the brands were not significant, but now in a two-way ANOVA we find that the brands are significant. This is because the within-brand variance of Exercise 4 is, for the most part, accounted for by the block factor in the two-way ANOVA, leaving a smaller SSE.

ANSWERS TO CHAPTER 6 QUIZ

1. different fertilizers

2. variance

3. normal; normal

4. between-samples; within-samples

5. have the same mean

6. 5% critical value of F

7. numerator; denominator

8. false; that's what we developed the ANOVA table for.

Regression and Correlation

Terms we'll be learning about in this chapter

bivariate data

correlation

dependent variable

function

independent variable

least squares

residuals

slope

7-1 REGRESSION

One of the most useful applications of inferential statistics is the ability to make predictions about the value of one variable from knowledge of other variables. For instance, a manufacturer might want to know how many units of his product he will sell if the unit price is $50, and how many units he will sell if his unit price is $55, $60, and so forth. Here the number of units sold is to be predicted by the unit price. In mathematical language, the number of units sold is a **function** of the unit price (that is, the number of units sold depends in some systematic way on the unit price of the goods). As a second example, a teacher might want to know how the scores on her recent test depend on the number of hours each student studied. Here the grade is a *function* of the number of hours studied.

In these examples, as in most real-life situations, the relationship between the variables, that is, between price and number of units sold, or between hours studied and test score, is some unknown (and unknowable) function. But for some range of values of the variables, *the function may be approximated by a linear function*, that is, one whose graph is a straight line. (See Figures 7-1 and 7-2.) That is the case we will be dealing with for most of our study of regression. We will be seeking the approximate linear relationship between the variables, assuming that we are given some data that relate the variables to one another. Another way of saying this is that we are going to *fit* a straight line to the data. Let's take a closer look at the exam-study example.

Suppose the teacher polls the class to find out the number of hours each student studied for the test. Then for each student she records the number of hours studied and the score that student received on the test, as follows:

Student	Hours (x_i)	Test Score (y_i)
Mark	10	70
Fred	15	78
Alice	15	85
Elaine	20	90

Notice that for each student she records *two* numbers, (x_i, y_i). We call this **bivariate** (two-variable) **data** because we know the values of two variables for each person. We actually have four items of bivariate data, corresponding to the four points (10, 70), (15, 78), (15, 85), and (20, 90). We can now plot these four points on a graph, as shown in Figure 7-3 (page 228). What we are looking for is a straight line that "fits" these data points. Now, of course, since the four points don't lie on a straight line, there is no straight line that will fit the data exactly. So we are forced to find the best-fitting line and accept the fact that it cannot fit exactly. For the moment, don't worry about how we determine which line is the best fitting. Suppose we have found it. Then we can use this line as the function that will *predict* what score another student will get on the test, based on the number of hours studied.

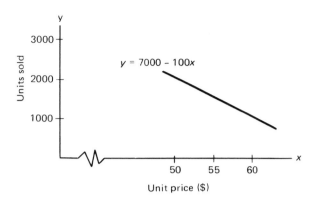

Figure 7-1

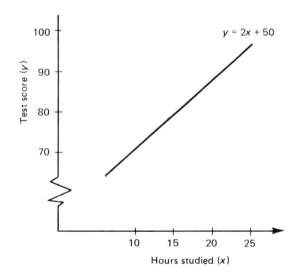

Figure 7-2

By the method of linear regression, which I'll explain shortly, we find the best-fitting straight line through the four data points to be the line with the equation $y = 2x + 50.75$. It looks like the line in Figure 7-4 (page 228). Now, based on this line, we can predict that a student who studies 12 hours will get a test score of

$$2(12) + 50.75 = 74.75.$$

And a student who studies 25 hours will get a score of

$$2(25) + 50.75 = 100.75,$$

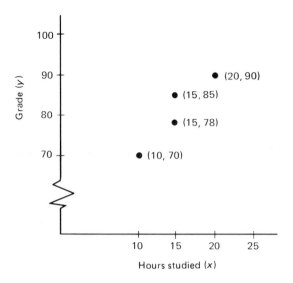

Figure 7-3

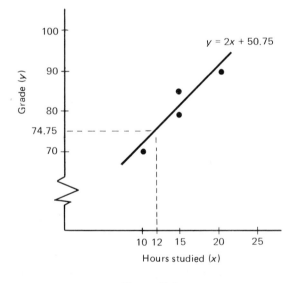

Figure 7-4

which points up the fallibility of the method. We have to keep in mind that the linear prediction function is only an approximation of the true state of things, and the predictions will only be accurate to the degree that the linear function is accurate. The linear function will approximate the data to the degree that the data are truly linearly related. The more linear the relationship in the data, the better the prediction. The less linear the relationship, the poorer the prediction. The degree of linear relationship in the data is measured by **correlation**. I'll tell you about correlation later in this chapter.

WHICH STRAIGHT LINE IS THE "BEST FIT"?

There are many possible criteria for choosing the "best fit." Some will make sense and others won't. Look at the examples in Figure 7-5 and you will see two lines that we would consider a good fit and one that we would consider a poor fit. Since we're developing a methodology here, we need a measure of "best fit" that once agreed upon will lead to the *same* straight line each time we start with the same data. Of the many possible criteria, the one we choose (for mathematical reasons, just like the one choice out of many for the variance as the measure of spread) is the line that has the least squared deviations of the data from the line, measured *vertically*. That is, vertical deviations are in the *y* direction. The variable *y* is the predicted or *depen-*

| Good fit | Good fit | Poor fit |

Figure 7-5

dent variable (in the case of our exams, the test score—predicted as a function of the number of hours studied). The hours studied, x, is the **independent variable**. The variable x (hours studied) is called the independent variable because we can "choose" the number of hours we study. The predicted test score, y, *depends* on x ($y = 2x + 50.75$) and so is the **dependent variable**. We choose as the best-fitting line, the line that has the smallest sum of squared deviations of the observed y-values (test scores) from the predicted y-values (predicted test scores) labeled y'. The deviations, $y - y'$, are called the **residuals**. In our example, the observed test scores are 70, 78, 85, and 90. If we use our line with equation $y = 2x + 50.75$ and use the hours studied as the values of x (the independent variable), we get predicted scores y'_i of 70.75, 80.75, 80.75, and 90.75, respectively (see Table 7-1). Of all possible straight-line fits to the data, no other line will give a sum of squared deviations of observed from predicted less than 26.75. This line is the best-fitting line according to the criterion of *least squares*. Thus, this method is called the **method of least squares**.

Table 7-1

Student	Hours (x_i)	Test Score (y_i)	Predicted Score $(y'_i = mx_i + b)$	Residuals $(y_i - y'_i)$	Squared Deviations $(y_i - y'_i)^2$
Mark	10	70	70.75	−0.75	0.5625
Fred	15	78	80.75	−2.75	7.5625
Alice	15	85	80.75	4.25	18.0625
Elaine	20	90	90.75	−0.75	0.5625
				$\Sigma(y_i - y'_i)^2 =$	26.75

HOW TO COMPUTE THE LEAST SQUARES LINE

Some calculators, such as the TI-55, have the calculation of the regression line built in. Simple calculators do not. However, the calculation of the regression equation is relatively simple. I'll show you how to make it on the data we have on the test scores. The procedure is the same for any set of data. You're going to be making the calculation with me, so get out your calculator.

In Table 7-2, I've entered the bivariate hours-score data for the four students. Notice that I've dropped the subscripts i on x_i and y_i. You fill in the remaining two columns by squaring the hours x and entering the results in the third column, marked x^2, and then multiplying hours x by test scores y and entering the results in the fourth column, marked xy. Then obtain column totals for all four columns and enter them in the appropriate places in the table.

Table 7-2

Hours (x)	Test Score (y)	x^2	xy
10	70	100	700
15	78		
15	85		
20	90		
Totals: $\Sigma x =$	$\Sigma y =$	$\Sigma x^2 =$	$\Sigma xy =$

Now that you've completed the entries in Table 7-2, compare your totals with my totals below:

Totals:	$\Sigma x = 60$	$\Sigma y = 323$	$\Sigma x^2 = 950$	$\Sigma xy = 4945$

Did you get the same results? If not, turn the page and take a look at Table 7-3, where I've reproduced the completed table. Compare Table 7-3 with your completed Table 7-2 and see where you went wrong.

Next, let n be the number of data points, which in this case is 4. Then calculate

$$m = \frac{n(\Sigma xy) - (\Sigma x)(\Sigma y)}{n(\Sigma x^2) - (\Sigma x)^2}, \tag{1}$$

$$b = \frac{(\Sigma x^2)(\Sigma y) - (\Sigma x)(\Sigma xy)}{n(\Sigma x^2) - (\Sigma x)^2}. \tag{2}$$

The regression line has the equation

$$y = mx + b,$$

where m and b are as calculated in (1) and (2). Note that in the calculation of m and b, the denominator is the same in both (1) and (2) and hence needs to be calculated only once. In the case of our data of scores and hours, we substitute the column totals into (1) and (2) to obtain

$$m = \frac{4(4945) - (60)(323)}{4(950) - (60)^2} = 2.00,$$

$$b = \frac{(950)(323) - (60)(4945)}{4(950) - (60)^2} = 50.75.$$

Table 7-3

	Hours (x)	Test Score (y)	x^2	xy
	10	70	100	700
	15	78	225	1170
	15	85	225	1275
	20	90	400	1800
Totals:	$\Sigma x = 60$	$\Sigma y = 323$	$\Sigma x^2 = 950$	$\Sigma xy = 4945$

Thus, the regression, or least squares line, for the given data is

$$y = 2x + 50.75.$$

The two numbers m and b that we calculated are called the **regression coefficients**. We can *predict*, based on our regression equation, that a student studying 12 hours will get a score of $2(12) + 50.75 = 74.75$ on the test.

THE SLOPE OF THE REGRESSION LINE

We know from analytic geometry that m, the coefficient of x in

$$y = mx + b,$$

gives the *slope* of the line in the (x, y)-plane. This slope has two significant features. The first of these is the sign. If the sign of the coefficient m is positive, then the equation predicts that with increasing x-values, there will be an increase in y-values. For the hours-score example, we expect that with increased number of hours studied (x-values), there will be an accompanying increase in test score (y-values). We find that this is indeed the case, as the coefficient $m = +2.00$. For the manufacturer's example, where with *increased* price we expect a *decrease* in total number of units sold, we would expect that the coefficient would be negative (as we shall see in Exercise 1).

The second important feature of the slope, or coefficient m, is that it gives the **marginal return** for increased study. The equation $y = 2x + 50.75$ predicts that each increased hour of study (x) will increase the test score (y) by 2 points. (Look at the equation. When you increase x by 1, by how much does y increase?) The *marginal return*, therefore, is 2.

EXERCISE 1   Suppose a manufacturer has the following information:

x (price in $):	50	55	60	70	80
y (units sold):	3000	2400	2000	1200	400

a. Complete the table below to find the regression equation for sales as a function of price that "best fits" the data.

b. Predict the number of units the manufacturer would sell if he priced his product at $40. At $90.

Data		x^2	xy
x	y		
Totals: $\Sigma x =$	$\Sigma y =$	$\Sigma x^2 =$	$\Sigma xy =$

n = _____

m = _____

b = _____

y = _____ x + _____

7-2 CORRELATION: MEASURING THE AMOUNT OF VARIANCE EXPLAINED BY LINEAR REGRESSION

Let's take another look at our study example. This time let's see how well the regression equation

$$y = 2x + 50.75$$

predicts the test scores. (See Table 7-4.) We know that the better the fit, the smaller the sum of squared residuals $\Sigma(y_i - y_i')^2$. If the data all lay on a straight line, we would obtain a perfect linear fit and the residuals would all be 0. One measure of how well the data are fit by a liner function, that is, how "linear" the data are, is the *correlation coefficient* of the data. The correlation coefficient measures the proportion of the total variance in the y-values (test scores) that is "explained" (predicted)

Table 7-4

Student	Hours (x)	Test Score (y)	Predicted Score (y')	Residual (y − y')
Mark	10	70	70.75	−0.75
Fred	15	78	80.75	−2.75
Alice	15	85	80.75	+4.25
Elaine	20	90	90.75	−0.75

by the x's—that is, the proportion of the total variance in the y-values that is in the predicted (y') values.

If we consider the residuals ($y_i - y_i'$) to be that portion of the test scores *not* accounted for by linear regression (think about it), then

$$\frac{\Sigma(y_i - y_i')^2}{\Sigma(y_i - \bar{y})^2} \tag{3}$$

represents the *proportion* of variance in the y-values (test scores) *not* accounted for by regression, and so

$$1 - \frac{\Sigma(y_i - y_i')^2}{\Sigma(y_i - \bar{y})^2} = r^2 \tag{4}$$

is the proportion of variance that *is* accounted for by linear regression. In (3) and (4), $\bar{y} = \Sigma y_i/n$, the mean of the observed y-values. [The denominator of the expression in (3) is n times the total variance in the y-values.] Then r, which is the square root of the expression r^2 in (4), is the **correlation coefficient**. Note that r is chosen $+$ or $-$ as the slope of the linear regression is $+$ or $-$ (given by the sign of the coefficient m). If the slope is positive, then y increases with increasing x's, as in the hours-score example. In this case we say that the data are **positively correlated**. If the slope is negative, then y decreases with increasing x's, as in the manufacturer's price-units example. In this case we say that the data are **negatively correlated**.

In our hours-score example,

$$\Sigma(y_i - \bar{y})^2 = (70 - 80.75)^2 + (78 - 80.75)^2 + (90 - 80.75)^2 + (85 - 80.75)^2$$
$$= 226.75,$$

and

$$\Sigma(y_i - y_i')^2 = (-0.75)^2 + (-2.75)^2 + (-0.75)^2 + (4.25)^2$$
$$= 26.75.$$

Then

$$r^2 = 1 - \frac{26.75}{226.75} = 0.8820287,$$

and

$$r = \pm 0.9391638.$$

Finally, $r = +0.9391638$ is the correlation coefficient, since the data are positively correlated (have positive slope). The value $r^2 \cong 0.88$ means that approximately 88% of the variance in the test scores is "explained" by the linear relationship between the hours studied and the test scores.

FAST FORMULA FOR COMPUTING CORRELATION COEFFICIENT

While (4) is a good way to define r, the correlation coefficient, in practice there is a simpler formula for computing it:

$$r = \frac{n(\Sigma xy) - (\Sigma x)(\Sigma y)}{\sqrt{n(\Sigma x^2) - (\Sigma x)^2}\ \sqrt{n(\Sigma y^2) - (\Sigma y)^2}} \tag{5}$$

All the values you need for (5) except Σy^2 can be found in Table 7-3.

EXERCISE 2

Compute r, the correlation coefficient, of the data of Exercise 1. _____

Is r positive or negative? _____

Explain. _____

_____

What percent of total variance in the y-values is explained by regression? _____

7-3 CORRELATION ANALYSIS: TESTING $\rho = 0$

We have to be somewhat careful not to give undue consideration to a large (positive or negative) value for r. Remember, the r we calculate is a statistic of the bivariate data, not of the underlying population correlation coefficient, which is a parameter. The population correlation coefficient measures the degree of linear relationship in the *bivariate population* of all hours-scores for *all* students, or all unit-price–units-sold figures for every possible price, and so on. We denote the population correlation coefficient ρ (the Greek letter rho), and, of course, r is just an estimator for ρ.

If we wish to test whether or not ρ is equal to 0, which is the common test for ρ, namely, that the (bivariate) population does not or does possess any linear correlation, we form the null hypothesis $\rho = 0$. Then under the null hypothesis and certain assumptions of normality, it is known that the statistic

$$t = \frac{r\sqrt{n-2}}{\sqrt{1-r^2}} \qquad (6)$$

has a t-distribution, $n - 2$ d.f.

Example 1

A social scientist obtains $n = 5$ observations of bivariate data, for which she calculates an $r = 0.85$. She rushes to her colleagues with excitement at the "new-found" linear relationship in the data. Her skeptical colleagues test the null hypothesis $\rho = 0$ by computing t as shown in (6):

$$t = \frac{0.85\sqrt{3}}{\sqrt{1-0.85^2}} = 2.79.$$

The 5% critical value for t (two-sided), $t_{0.025}$, 3 d.f., is 3.182. Since 2.79 does not exceed this, the social scientist *cannot reject* the null hypothesis that the data come from a bivariate population with no linear correlation, at the 5% level. We say "the value $r = 0.85$ is *not significant* at the 5% level." ∎

EXERCISE 3

Is an observed correlation of -0.78 significant at the 1% level, for $n = 15$ (15 data points)? _____

7-4 MULTIPLE LINEAR REGRESSION

Suppose, in our hours-score example, that we not only know the number of hours each student studied, but also know the grade that each student received in the previous course. If we use the number of hours *and* the previous course grades to predict the test scores, we should be able to make more accurate predictions than we did before. Again, we will be predicting the test score as a linear function, this time a linear function of the hours studied *and* the previous course grades. Such a linear function looks like this:

$$y = ax + bu + c, \qquad (7)$$

where a, b, and c are constants, $x =$ the number of hours studied, and $u =$ the previous course grade. This is a linear function of two variables, x and u. Earlier we predicted y from the one variable x. Now we predict y from x and u.

This time we are presented with trivariate data (three variables): x, u, and y (see Table 7-5). If you look carefully at the data, you can see that we can partially explain why Alice scored better on the test (85) than Fred did (78), even though they both studied the same number of hours (15). Alice outscored Fred in the previous course, which we would expect to predict the grades on the test to a certain extent.

Table 7-5

Student	Hours (x)	Previous Course Grade (u)	Test Score (y)
Mark	10	75	70
Fred	15	75	78
Alice	15	90	85
Elaine	20	80	90

Recall that when we did linear regression of the grades (y) on the hours studied (x), we got the linear equation

$$y = 2x + 50.75,$$

which gave the linear relationship between the x-values and the y-values. Our goal now is to obtain an equation like (7), where the regression coefficients a, b, and c are replaced by numbers. We construct a table, like we did in Table 7-1 for the *one-variable regression* of y on x. This time it's for y on x and u, and there are more columns (see Table 7-6).

I've already entered the trivariate data in Table 7-6. Take out your calculator and complete the entries. Look back at Table 7-3 if necessary to see how we did it.

Table 7-6

x	u	y	x^2	u^2	ux	uy	xy
10	75	70					
15	75	78					
15	90	85					
20	80	90					
Totals: $\Sigma x =$	$\Sigma u =$	$\Sigma y =$	$\Sigma x^2 =$	$\Sigma u^2 =$	$\Sigma ux =$	$\Sigma uy =$	$\Sigma xy =$

After you've filled in the table completely, compute the column totals and write them at the bottom of each column. These are the numbers we'll need to compute the regression coefficients a, b, and c. Compare your column totals with mine:

$\Sigma x = 60$	$\Sigma u = 320$	$\Sigma y = 323$	$\Sigma x^2 = 950$	$\Sigma u^2 = 25{,}750$	$\Sigma ux = 4{,}825$	$\Sigma uy = 25{,}950$	$\Sigma xy = 4{,}945$

If you didn't get these totals, compare your Table 7-6 with my Table 7-7.

Table 7-7

x	u	y	x^2	u^2	ux	uy	xy
10	75	70	100	5,625	750	5,250	700
15	75	78	225	5,625	1,125	5,850	1,170
15	90	85	225	8,100	1,350	7,650	1,275
20	80	90	400	6,400	1,600	7,200	1,800
Totals: $\Sigma x = 60$	$\Sigma u = 320$	$\Sigma y = 323$	$\Sigma x^2 = 950$	$\Sigma u^2 = 25{,}750$	$\Sigma ux = 4{,}825$	$\Sigma uy = 25{,}950$	$\Sigma xy = 4{,}945$

Now we enter these totals in the so-called *normal equations* below:

$$(\Sigma y) = \quad nc + (\Sigma x)a + (\Sigma u)b, \tag{8}$$

$$(\Sigma xy) = (\Sigma x)c + (\Sigma x^2)a + (\Sigma ux)b, \tag{9}$$

$$(\Sigma uy) = (\Sigma u)c + (\Sigma uy)a + (\Sigma u^2)b. \tag{10}$$

Replacing the quantities in parentheses with the appropriate column totals, and recalling that $n = 4$ (four observations, or data points), we get

$$323 = \quad 4c + \quad 60a + \quad 320b, \tag{11}$$

$$4945 = \quad 60c + \quad 950a + \quad 4825b, \tag{12}$$

$$25{,}950 = 320c + 4825a + 25{,}750b. \tag{13}$$

Now that we have three linear equations in three unknowns, a, b, and c, we can use high school algebra (the method of Gaussian elimination) to solve these equations. Multiply equation (11) by 15 and then subtract it from equation (12). This will *eliminate* the varable c from the resulting equation, which appears below as equation (15). Similarly, multiply equation (11) by 80 and then subtract it from equation (13). This will eliminate the variable c from the resulting equation (16). We get

$$323 = 4c + 60a + 320b, \qquad (11) = (14)$$

$$100 = + 50a + 25b, \qquad (15)$$

$$110 = + 25a + 150b. \qquad (16)$$

Next we attempt to eliminate the variable a from the third equation, (16), the same way we eliminated c from equations (15) and (16). If we multiply equation (16) by 2 and subtract it from equation (15), the resulting equation will contain no variable a and will become our new third equation:

$$323 = 4c + 60a + 320b. \qquad (14) = (17)$$

$$100 = + 50a + 25b, \qquad (15) = (18)$$

$$-120 =   - 275b. \qquad (19)$$

Notice that by this systematic method (which would work as well on four equations in four unknowns, and so forth), we have reduced our equations to "upper triangular form"; now we can easily solve them from the bottom up. From (19), we get

$$b = \frac{-120}{-275} = 0.436. \qquad (20)$$

Now that we know b, from (18) we get

$$100 = 50a + 25(0.436)$$

$$a = \frac{100 - 10.9}{50} = 1.782. \qquad (21)$$

Now that we know a and b, from (17) we get

$$323 = 4c + 60(1.782) + 320(0.436)$$

$$c = \frac{323 - 60(1.782) - 320(0.436)}{4} = 19.14. \qquad (22)$$

So you can see that once we reduce the linear system of equations to upper triangular form like (17), (18), and (19)—even if we have 10 equations!—it's a rather easy computation to solve for the unknowns. Equation (7) becomes

$$y = 1.782x + 0.436u + 19.14. \qquad (23)$$

Table 7-8

		Predicted (y')	Observed (y)	Residual (y − y')
Mark:	$1.782(10) + 0.436(75) + 19.14 =$	69.66	70	0.34
Fred:	$1.782(15) + 0.436(75) + 19.14 =$	78.57	78	−0.57
Alice:	$1.782(15) + 0.436(90) + 19.14 =$	85.11	85	−0.11
Elaine:	$1.782(20) + 0.436(80) + 19.14 =$	89.66	90	0.34

Equation (23) is our regression equation. You can see how well (23) *predicts* the grades (y) in Table 7-8. We can still use (3) and (4) to determine how much of the variance in the observed *y*-values is accounted for by regression:

$$\Sigma(y_i - \bar{y})^2 = 226.75,$$

$$\Sigma(y_i - y_i')^2 = (0.34)^2 + (-0.57)^2 + (-0.11)^2 + (0.34)^2$$

$$= 0.5682.$$

Then

$$r^2 = 1 - \frac{0.5682}{226.75} = 0.99749.$$

And so, regression accounts for about 99.75% of all the variance in the *y*-values (test scores). We can conclude that our linear model for predicting test scores from hours studied and the former course grade is pretty good.

EXERCISE 4

In order to study the effect of IQ and age on income, fit an equation of the form $y = ax + bu + c$ to the data below, where x = age and u = IQ. Then predict what income a 42-year-old with an IQ of 115 will earn.

Age (x)	IQ (u)	Income (y)
38	116	32,700
46	100	28,300
39	120	36,500
43	108	31,800
32	116	26,900

LINEAR REGRESSION WITH MORE VARIABLES

When we did linear regression of y on x (bivariate), we ended up with two (normal) equations in two unknowns. We didn't bother to write out the equations, but just gave the solution for m and b, equations (1) and (2). For three variables (trivariate data), we got three equations in three unknowns, and I showed you how to solve them. For more variables we get more equations, but the method of solving them is the same. It's just a bit more complicated to write down the normal equations. In the back of the book, I give a BASIC program that will calculate the regression equation for you, for as many variables as you have. But you'll need a computer to use the program. If, instead, you want to use your calculator for a four- or five-variable linear regression, it will take you quite a bit of time. [See N. R. Draper and H. Smith, *Applied Regression Analysis* (New York: John Wiley & Sons, 1966) for details.] For larger numbers of equations, you're better off using the computer program.

7-5 FITTING TO CURVES OTHER THAN STRAIGHT LINES

There are many situations where we don't want to fit the data to a straight line. For instance, we would expect growth curves to be exponential curves rather than straight lines (see Figure 7-6). This curve describes the growth of a colony of bacteria over time. A similar curve describes the sales over time of a newly established company.

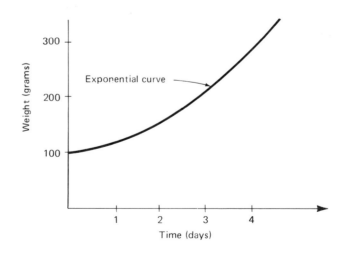

Figure 7-6

The method of linear regression applies when we want to fit such curves to data if we still measure "fit" by squared deviations. We transform the curve (by whatever means) to a straight line, *transforming the given bivariate data at the same time*. Then we fit a straight line to the data by linear regression, and then transform back again.

Example 2

We own a small company, and much of our business comes from referral from previous customers. So the more customers we serve, the more business we expect to get. This situation is typically described by an exponential growth curve. Suppose the number of sales per year for each of the first six years is as follows:

Year	Sales
1	109
2	146
3	235
4	350
5	575
6	852

Let's attempt to fit an exponential growth curve

$$y = A \cdot 10^{Bx}, \tag{24}$$

where A and B are constants to be found. Taking logs of both sides of (24) we get

$$\log y = \log A + Bx$$
$$= Bx + \log A. \tag{25}$$

Notice that (25) is a *linear equation* in x and $\log y$. We transform the data, accordingly, to $(x, \log y)$-values.

$$\text{Data:} \quad (1, 2.04) \quad (2, 2.16) \quad (3, 2.37)$$
$$(4, 2.54) \quad (5, 2.76) \quad (6, 2.93)$$

We construct the usual table, but in the variables x and $\log y$ (see Table 7-9). We get

$$m = \frac{6(55.01) - (21)(14.80)}{6(91) - (21)^2} = \frac{19.26}{105} = 0.18,$$

$$b = \frac{(91)(14.80) - (21)(55.01)}{6(91) - (21)^2} = \frac{191.59}{105} = 1.82.$$

Table 7-9

x	$\log y$	x^2	$x \cdot \log y$
1	2.04	1	2.04
2	2.16	4	4.32
3	2.37	9	7.11
4	2.54	16	10.16
5	2.76	25	13.80
6	2.93	36	17.58
Totals: 21	14.80	91	55.01

The linear equation (25) is of the form

$$y = mx + b,$$

where $m = B$ and $b = \log A$. So having found m and b, (25) becomes

$$\log y = 0.18x + 1.82. \tag{26}$$

Now if we exponentiate both sides of (26),

$$10^{\log y} = 10^{(0.18x + 1.82)} \tag{27}$$

or

$$y = 10^{0.18x} \cdot 10^{1.82} \tag{28}$$

or

$$y = (66.07)10^{0.18x} \tag{29}$$

or

$$y = 66.07(e^{2.30})^{0.18x} \tag{30}$$

or

$$y = 66.07e^{0.41x} \tag{31}$$

Formulas (29) and (31) are the usual form for this equation. Using (29), we can predict our sales for the seventh year:

$$y = (66.07)(10^{0.18(7)}) = 1202.3$$

predicted sales. [See N. R. Draper and H. Smith, *Applied Regression Analysis* (New York: John Wiley & Sons, 1966).] ∎

PROBLEM SET A   1. The following bivariate data relate the values of the variables x and y. Assuming that there exists a linear relationship between x and y, compute the equation of the regression line.

x	10	15	15	20
y	72	80	86	91

2. The following bivariate data relate the values of the variables x and y. Assuming that there exists a linear relationship between x and y, compute the equation of the regression line.

x	50	55	60	70	80
y	3000	2400	2000	1200	402

3. The following bivariate data relate the values of the variables x and y. Assuming that there exists a linear relationship between x and y, compute the equation of the regression line.

x	1390	1450	1570	1700	1810	1940	2080	2330
y	1383	1374	1466	1694	1771	1911	2005	2040

4. The following bivariate data relate the values of the variables x and y. Assuming that there exists a linear relationship between x and y, compute the equation of the regression line.

x	2	3	5	6	8	8	9	10	10	12
y	12	16	26	31	40	43	47	50	56	62

5. The following bivariate data relate the values of the variables x and y. Assuming that there exists a linear relationship between x and y, compute the equation of the regression line.

x	1	2	3	4	5
y	3	2	4	6	7

6. The following bivariate data relate the values of the variables x and y. Assuming that there exists a linear relationship between x and y, compute the equation of the regression line.

x	7	7	5	3	4
y	22	20	11	6	13

7. The following bivariate data relate the values of the variables x and y. Assuming that there exists a linear relationship between x and y, compute the equation of the regression line.

x	1	2	3	4	5	6	7	8
y	25	32	30	28	32	36	37	39

PROBLEM SET B

1. The following bivariate data relate the values of the variables x and y. Compute the correlation coefficient of the given data.

x	10	15	15	20
y	72	79	85	91

2. The following bivariate data relate the values of the variables x and y. Compute the correlation coefficient of the given data.

x	50	55	60	70	80
y	3000	2400	2002	1200	402

3. The following bivariate data relate the values of the variables x and y. Compute the correlation coefficient of the given data.

x	1390	1450	1570	1700	1810	1940	2080	2330
y	1383	1373	1464	1694	1773	1913	2006	2042

4. The following bivariate data relate the values of the variables x and y. Compute the correlation coefficient of the given data.

x	2	3	5	6	8	8	9	10	10	12
y	11	17	26	30	40	42	45	52	57	62

5. The following bivariate data relate the values of the variables x and y. Compute the correlation coefficient of the given data.

x	1	2	3	4	5
y	2	3	5	5	6

6. The following bivariate data relate the values of the variables x and y. Compute the correlation coefficient of the given data.

x	7	7	5	3	4
y	20	18	12	7	13

7. The following bivariate data relate the values of the variables x and y. Compute the correlation coefficient of the given data.

x	1	2	3	4	5	6	7	8
y	27	33	31	29	34	34	36	37

CHAPTER 7 QUIZ

1. Regression is a statistical technique that allows us to _____ the value of one variable from knowledge of another variable.

2. The given data in one-variable regression are numbers that occur in _____ .

   These represent _____ on the graph of y versus x.

3. Regression allows us to fit a _____ to the given points.

4. The choice of which line is the best-fitting straight line is determined by the criterion of _____ .

5. The slope of the regression line is given by _____

    _____ in $y = mx + b$.

6. The differences between observed y-values and predicted y-values are called

    the _____ .

7. The _____ measures

    the degree of linearity in the data. It is measured by the statistic _____ .

ANSWERS TO EXERCISES

1.

x	y	x^2	xy
50	3,000	2,500	150,000
55	2,400	3,025	132,000
60	2,000	3,600	120,000
70	1,200	4,900	84,000
80	400	6,400	32,000
Totals: 315	9,000	20,425	518,000

$$m = \frac{5(\Sigma xy) - (\Sigma x)(\Sigma y)}{5(\Sigma x^2) - (\Sigma x)^2} = \frac{-245,000}{2900} = -84.48;$$

$$b = \frac{(\Sigma x^2)(\Sigma y) - (\Sigma x)(\Sigma xy)}{2900} = \frac{20,655,000}{2900} = 7122.41;$$

$$y = -84.48x + 7122.41.$$

At \$40: $y = -84.48(40) + 7122.41 = 3743$ units.
At \$90: $y = -84.48(90) + 7122.41 = -481$ units.

(*Note*: \$90 is undoubtedly out of the range where the linear function is a good predictor, as it predicts a negative number of units sold.)

2. $r = -0.99755$. We expect r to be negative as the number of sales *decreases* as the price increases (slope is negative).

   $r^2 = 0.995$; so regression accounts for 99.5% of all the variance in the y-values.

3. $t = \dfrac{(-0.78)\sqrt{15 - 2}}{\sqrt{1 - (-0.78)^2}} = -4.494.$

   With 13 d.f., critical value is $t_{0.005} = 3.012$. Since -4.494 is more negative than -3.012, $r = -0.78$ is significant at the 0.01 level (note two-tailed: 0.005 in each tail).

4.

	x	u	y	x^2	u^2	ux	uy	xy
	38	116	32,700	1,444	13,456	4,408	3,793,200	1,242,600
	46	100	28,300	2,116	10,000	4,600	2,830,000	1,301,800
	39	120	36,500	1,521	14,400	4,680	4,380,000	1,423,500
	43	108	31,800	1,849	11,664	4,644	3,434,400	1,367,400
	32	116	26,900	1,024	13,456	3,712	3,120,400	860,800
Totals:	$\Sigma x =$ 198	$\Sigma u =$ 560	$\Sigma y =$ 156,200	$\Sigma x^2 =$ 7,954	$\Sigma u^2 =$ 62,976	$\Sigma ux =$ 22,044	$\Sigma uy =$ 17,558,000	$\Sigma xy =$ 6,196,100

$$156,200 = 5c + 198a + 560b \qquad (1)$$

$$6,196,100 = 198c + 7954a + 22,044b \qquad (2)$$

$$17,558,000 = 560c + 22,044a + 62,976b \qquad (3)$$

Multiply equation (1) by 198/5 and subtract it from equation (2):

$$10,580 = 113.2a - 132b. \qquad (4)$$

Multiply equation (1) by 560/5 and subtract it from equation (3):

$$63,600 = - 132a + 256b. \qquad (5)$$

Now use equations (4) and (5) to eliminate $-132a$ from equation (5). Multiply equation (4) by 132/113.2 and *add* it to (5):

$$75,937.102 = 102.0777b. \qquad (6)$$

Equations (1), (4), and (6) are in upper triangular form, and (6) gives $b = 743.914$. Substituting this into (4), $a = 960.925$. Then substituting into (1), $c = -90,131$. The equation, then, is

$$y = 960.925x + 743.914u - 90,131.$$

Finally, if $x = 42$ and $u = 115$, then $y = 35,778$.

ANSWERS TO CHAPTER 7 QUIZ

1. predict, or infer

2. pairs; points

3. straight line

4. least squares

5. m; the coefficient of x

6. residuals

7. correlation coefficient; r

Nonparametric Statistics

Terms we'll be learning about in this chapter

nonparametric tests

paired data

rank-sum

sign test

8-1 INTRODUCTION

Up to now, many of the statistical tests we have discussed depended on one or more assumptions. We assumed that one or more populations were normal or approximately normal, or that they had a certain standard deviation, or that they all had the same standard deviation, and so on. We call such assumptions *parametric* if they involve parameters (like sigma) of a given population. In this chapter we will study nonparametric statistical tests and distribution-free tests. The distribution-free tests are those that make no assumption about the shape of the distribution of a given population; and of course the nonparametric tests make no assumptions about the parameters. We'll call both kinds of tests **nonparametric tests**.

8-2 THE ONE-SAMPLE SIGN TEST

With the one-sample **sign test** you can test any population or distribution's *median* $m = m_0$ (some given value) against either a two-sided alternative, $m \neq m_0$ or a one-sided alternative, say $m > m_0$. The sign test works as follows. Say you wish to test $m = 5$ against the alternative $m \neq 5$. If the median is 5, then we expect that half the elements drawn randomly from the population will be greater than 5, and half will be less than 5. So, draw out a sample of size 15, say, from the population, and assign $+$ to each element greater than 5, assign $-$ (minus) to each element less than 5, and throw away any 5s you observe. Then if the hypothesis $m = 5$ is correct, each $+$ or $-$ can be regarded as the outcome of a *coin toss* with $p = \frac{1}{2}$. Now we are reduced to testing the hypothesis: $p = \frac{1}{2}$ from a sequence of 15 tosses of a coin, recording $+$ for heads and $-$ for tails. This is just the binomial distribution, $n = 15$, $p = \frac{1}{2}$. (For small values of n, we would look this up in a table of the binomial distribution; for large values of n we can use the normal approximation to the binomial.)

AN EXAMPLE OF THE SIGN TEST

The following are breaking strengths of a certain size beam, tested 15 times:

6.1	6.1	5.9	6.0	6.2	6.2	6.3	6.1
6.1	5.9	6.1	6.3	6.1	5.9	6.2	

Let's test the hypothesis that the median is 6.0. Using the sign test, we compare the 15 observations to 6.0 and record $+$ for any larger than 6.0 and $-$ for any less than 6.0. We discard the one observation of 6.0, leaving 14 observations:

$+$	$+$	$-$		$+$	$+$	$+$	$+$
$+$	$-$	$+$	$+$	$+$	$-$	$+$

We observe 11 pluses and 3 minuses. Now we look in Table 7a, "Critical Values of the Binomial Distribution," in the back of the book. This table tells us, for each sample size from 5 to 15, how many pluses and minuses we need to observe in order to reject the hypothesis $p = \frac{1}{2}$ at either the 0.05 level or the 0.01 level of significance. (If $n > 15$, part b of the table shows how to compute a critical value for both the 0.05 and 0.01 levels.) We find that the *two-tailed* 5% critical value for $n = 14$ is 12 or 2, meaning that the probability of observing 2 or fewer or 12 or more in 14 binomial trials with $p = \frac{1}{2}$ is less than 5% (actually 0.02). Observing 11 pluses, we cannot reject $p = \frac{1}{2}$. So we do not reject the hypothesis that the median is 6.0. If, however, the fourteenth observation of 5.9 had been 6.2, say, then we would have observed 12 pluses and 2 minuses, and by use of our table of critical values of the binomial, we would reject the hypothesis that the median is 6.0, at the 0.05 level.

We can also use the same test for the mean of the population instead of the median, if we know that the distribution of the population is *symmetric*, because then the mean and the median are the same.

EXERCISE 1

We took the bus to work on eight different occasions. The travel times were 59, 48, 46, 55, 61, 47, 48, and 58 minutes. Test the hypothesis that the median travel time to work on the bus is *more than* 60 minutes

a. at the 0.01 level of significance.

_____

b. at the 0.05 level of significance.

_____

EXERCISE 2

For the same data as in Exercise 1, test the hypothesis that the median time to work *equals* 60 minutes

a. at the 0.05 level.

_____

b. at the 0.01 level.

_____

EXERCISE 3

The breaking strengths of a sample of 18 boxes were tested in a laboratory. The results were recorded (in pounds) as follows:

58	47	53	56	58	60	56	49	54
57	51	54	56	50	56	55	59	57

Do these data support a claim that the median breaking strength of this kind of box is 55 pounds? Use the sign test. ⎯⎯⎯⎯⎯⎯⎯⎯⎯⎯⎯⎯⎯⎯⎯⎯⎯

EXERCISE 4

For the data of Exercise 3, test the hypothesis that the median breaking strength is 55 pounds *or more*.

EXERCISE 5

A psychologist is testing a subject who claims to have ESP. The subject is to predict whether the next flip of a coin will be heads or tails. The psychologist then records whether the subject is right or wrong. The subject is correct on 57 out of 100 flips. Do the data indicate that the subject has ESP? (Perform a one-tailed test, 57 pluses, $n = 100$. This is one-tailed because a deviation of fewer correct answers than expected by chance would *not* be considered support for ESP.)

8-3 THE TWO-SAMPLE SIGN TEST

We can use the same test, the sign test, to determine whether the *median* of one population is greater than the median of another population. As with our original discussion of hypothesis testing, we can test median *A* greater than median *B*, or median *A* not equal to median *B*, depending on whether we make a one-tailed or two-tailed test. We'll give examples of each.

PAIRED DATA EXAMPLE

Two Olympic judges are being studied. It's thought that judge *A* gives higher scores than judge *B*. Here are the scores that 15 divers received from both judges:

Judge *A*:	5.9	6.0	6.2	6.3	6.2	6.1	5.9	5.8
	5.7	5.8	6.2	6.2	6.3	6.2	6.1	
Judge *B*:	6.0	5.9	5.9	5.9	5.9	6.0	5.7	5.7
	5.7	5.7	6.0	6.0	6.2	6.3	6.0	

In this example, the data are **paired** because the two judges are each rating the same set of divers. So we compare directly and replace each pair of numbers by a + if judge *A* scored higher than judge *B* and by a − if judge *A* scored lower than judge *B*. (It wouldn't matter if we reversed the pluses and minuses here.) Now, since the

Olympic Committee is investigating the question (testing the hypothesis) of whether judge *A* gives higher scores than judge *B*, we form the alternative hypothesis that judge *A* gives equal or lower scores than judge *B*, leading to a *one-tailed test*. We toss out the one case where the numbers are equal, leaving 12 pluses and 2 minuses:

$$- \quad + \quad + \quad + \quad + \quad + \quad + \quad +$$
$$+ \quad + \quad + \quad + \quad - \quad +$$

From Table 7a, we see that for one-tailed tests, we can reject the hypothesis $p = \frac{1}{2}$ at the 0.05 level if we have 11 or more pluses, $n = 14$. At the 0.01 level we can reject the hypothesis $p = \frac{1}{2}$ if we have 12 or more pluses, $n = 14$. Since we have 12 pluses and 2 minuses, we can reject the hypothesis that judge *A* gives equal or lower scores than judge *B* and conclude that judge *A* gives higher scores than judge *B* at the 0.01 significance level.

UNPAIRED DATA EXAMPLE

In contrast with our previous example, let's suppose that the 15 scores of judge *A* and 15 scores of judge *B* were for 30 *different* divers. Since the 30 scores are now *unpaired*, we must pair the scores *randomly*. To accomplish this, we look up 15 random numbers in our table of random numbers (Table 8, at the back of the book) and associate to each of the 15 scores of judge *A* one of these random numbers, as follows:

5.9	(17)
6.0	(41)
6.2	(79)
6.3	(32)
6.2	(33)
6.1	(98)
5.9	(44)
5.8	(51)
5.7	(21)
5.8	(64)
6.2	(94)
6.2	(65)
6.3	(42)
6.2	(24)
6.1	(23)

We got these random numbers by taking the first 2 digits of each number from the first column of random numbers in Table 8. We then *sort* the 15 scores according to the random numbers associated, taking as the first score the score 5.9, which has the lowest associated random number, 17. Our second score is 5.7, because its associated random number is 21, the next lowest. We continue in this way until we have rearranged all 15 of judge *A*'s scores "in random order." The new order is as follows:

5.9	(17)
5.7	(21)
6.1	(23)
6.2	(24)
6.3	(32)
6.2	(33)
6.0	(41)
6.3	(42)
5.9	(44)
5.8	(51)
5.8	(64)
6.2	(65)
6.2	(79)
6.2	(94)
6.1	(98)

And if two of the two-digit numbers were the same, we would take a third digit from the table for each, to decide which came first.

Now that we have "shuffled" the 15 scores of judge A into "random" order, we pair them with the scores from judge B. There is no need to "randomize" the order of the scores from judge B. *Randomizing one of the two lists is sufficient.* Now we pair the 15 scores from judge A with the 15 scores from judge B:

Judge A:	5.9	5.7	6.1	6.2	6.3	6.2	6.0	6.3
	5.9	5.8	5.8	6.2	6.2	6.2	6.1	
Judge B:	6.0	5.9	5.9	5.9	5.9	6.0	5.7	5.7
	5.7	5.7	6.0	6.0	6.2	6.3	6.0	
	−	−	+	+	+	+	+	+
	+	+	−	+	0	−	+	

After comparing judge A and judge B scores, assigning pluses and minuses, and throwing out ties (labeled 0), we observe 10 pluses and 4 minuses. Looking in Table 7a under $n = 14$ (we threw out 1 tie from our 15 pairs), we observe that we cannot reject the hypothesis $p = \frac{1}{2}$ at either the 0.05 level or the 0.01 level of significance (to reject one-tailed at the 0.05 level with $n = 14$ requires 11 or more pluses or minuses, and we have 10).

Notice that the same data, if paired, lead to the conclusion that one judge's scores were higher than the other at the 0.01 level. But if we shuffle the data, as we must if the data are unpaired, then we "lose information" in the process and cannot reach the same conclusion.

IF ONE SAMPLE IS LARGER THAN THE OTHER

This is necessarily a case of unpaired data, for if the data came in pairs, there would be exactly as many in one sample as the other. In this case, we take the *longer* list and assign random numbers to each element, as we just did for the unpaired data. Say $n = 20$ for the longer list and $n = 15$ for the shorter list. We assign random numbers to each of the 20 numbers of the longer list and sort the 20 numbers in increasing order of the associated random numbers. Then we take the first 15 numbers from the longer list and pair them with the 15 elements of the shorter list. The remaining 5 data elements are discarded. Because we associated random numbers with the 20 elements of the first list and then chose those with the 15 smallest random numbers, we are in effect picking a "random" subset of 15 out of the list of 20 before we do our pairing and comparison with the second list. This is a legitimate random sampling procedure, one that should be used whenever some of the elements from a collection of data are to be discarded (or selected).

EXERCISE 6   Powerhouse Golf Ball Company has developed a new long-distance golf ball. Fifteen golfers are asked to play a round of golf with the new golf ball and a round with Powerhouse's old golf ball. The results are recorded below:

New ball:	75	77	79	82	83	85	89	79
	80	81	85	87	83	81	77	

Old ball:	77	78	79	85	82	87	91	81
	82	81	88	88	84	82	78	

Test the hypothesis that the new ball results in lower golf scores than the old ball, at the 0.01 level of significance.

EXERCISE 7   Seven rats are fed diet *A*. Ten rats are fed diet *B*. The weight gains (in grams) for each group are recorded as follows:

Diet *A*:	15	21	17	19	23	24	19			
Diet *B*:	21	25	27	29	25	24	27	29	28	26

Use the unpaired comparison sign test to test the hypothesis that there is no difference in (median) weight gain resulting from the two diets, at the 0.05 level of significance.

8-4 TWO-POPULATION RANK-SUM: THE MANN–WHITNEY (*U*-) TEST

The sign test is crude in that it regards only which of the pair of numbers is larger, not the actual sizes of the numbers. Also, when the data have unequal sample sizes, we must discard data to use the sign test. An improvement over the sign test to compare two populations is the Mann–Whitney, or **rank-sum**, test. With this test, we take samples from two populations, say, *A* and *B*. The sample sizes need not be the same. We pool the data from the two samples into one list, which we then arrange in increasing order. We then look to see if the elements of the two samples *A* and *B* are "sprinkled" throughout the ordered list, or whether most of the elements of *A* come "ahead" of most of the elements of *B*, or vice versa. If the two populations are identical, we do not expect most of the elements of one sample to come "ahead" of the elements of the other sample, and we reject the hypothesis that the two populations are identical if they do. The *U*-statistic measures the degree of mixing of the two samples.

Example 1

Sample A:	1.0	1.1	1.2	1.4	2.2	2.3	2.5	3.0	3.3	3.7
Sample B:	4.1	4.4	4.9	5.0	5.1	5.3	6.2	6.6	6.8	6.9

Pooled and ordered:	1.0	1.1	1.2	1.4	2.2	2.3	2.5	3.0	3.3	3.7
	4.1	4.4	4.9	5.0	5.1	5.3	6.2	6.6	6.8	6.9

Comes from:	A	A	A	A	A	A	A	A	A	A
	B	B	B	B	B	B	B	B	B	B

In this case, all the elements of sample A come "ahead" of all the elements of sample B. If the two populations from which these samples come were identical, the likelihood that this would occur is very small. We will soon compute a value of the U-statistic and reject the null hypothesis that the populations are identical, concluding that the populations are not identical. ■

Example 2

Sample A:	1.0	1.1	2.1	2.2	2.7	4.0
Sample B:	0.7	1.5	2.3	3.5	3.6	

Pooled and ordered:	0.7	1.0	1.1	1.5	2.1	2.2
	2.3	2.7	3.5	3.6	4.0	

Comes from:	B	A	A	B	A	A
	B	A	B	B	A	

We will soon compute a value of the U-statistic and conclude that we cannot reject the hypothesis that the populations are not identical. There is too much mixing of elements of population A and population B. ■

COMPUTING THE *U*-STATISTIC

Example 1 Revisited

To compute the U-statistic, we indicate the rank, or position, of each element of the pooled list along with the sample it came from:

Data:	1.0	1.1	1.2	1.4	2.2	2.3	2.5	3.0	3.3	3.7
Rank:	1	2	3	4	5	6	7	8	9	10
Sample:	A	A	A	A	A	A	A	A	A	A

Data:	4.1	4.4	4.9	5.0	5.1	5.3	6.2	6.6	6.8	6.9
Rank:	11	12	13	14	15	16	17	18	19	20
Sample:	B	B	B	B	B	B	B	B	B	B

We add up the ranks of all the elements of one or the other sample. It doesn't matter which we choose. Since it's easier to add up the smaller numbers, we add up the ranks of sample A: $1 + 2 + 3 + 4 + 5 + 6 + 7 + 8 + 9 + 10 = 55$. We call this value R_A. Then R_B, the sum of the ranks of sample B, can be computed by the formula

$$R_B = \frac{n(n + 1)}{2} - R_A, \tag{1}$$

where n is the total number of numbers, in this case 20 (10 in sample A and 10 in sample B). Note that if we had first obtained R_B, we would now use the formula

$$R_A = \frac{n(n + 1)}{2} - R_B. \tag{2}$$

By formula (1),

$$R_B = \tfrac{1}{2}(20)(21) - 55 = 155.$$

Then we compute

$$U_A = n_A n_B + \frac{(n_A)(n_A + 1)}{2} - R_A, \tag{3}$$

where n_A = size of sample A, and

$$U_B = n_A n_B + \frac{(n_B)(n_B + 1)}{2} - R_B, \tag{4}$$

where n_B = size of sample B. We then set

$$U = \text{The } \textit{smaller} \text{ of } U_A \text{ and } U_B. \tag{5}$$

For Example 1,

$$U_A = (10)(10) + \frac{(10)(11)}{2} - 55 = 100,$$

$$U_B = (10)(10) + \frac{(10)(11)}{2} - 155 = 0,$$

$$U = (\text{Smaller of 100 and 0}) = 0.$$

When n_A and n_B are both less than 11, we use a table of critical values of U, Table 9 (at the back of the book). If the value we have computed for U is *smaller* than or equal to the critical value for U in the table, we reject the null hypothesis (that the two populations from which the samples come are identical). If the value of U is greater than the critical value in the table, then we cannot reject the null hypothesis. So, looking in Table 9, we see that for $n_A = n_B = 10$, we let $n_1 = n_A = 10$ and $n_2 = n_B = 10$, and the 0.01 critical value for U is 16. Since $U = 0$ is less than 16, we reject the null hypothesis (that the populations are identical) at the 0.01 level of significance.

When one of the sample size is larger than 11, we can test U without use of Table 9, because U is then approximately normally distributed with mean

$$\mu_U = \frac{n_A n_B}{2}$$

and standard deviation

$$\sigma_U = \sqrt{\frac{n_A n_B (n_A + n_B + 1)}{12}}.$$

So we compute a z-score,

$$z = \frac{\mu_U - U - \frac{1}{2}}{\sigma_U},$$

and see whether this score is significant at whatever our level of significance, usually 0.01 or 0.05. Because z will be large whether U_A or U_B is small, we make a two-tailed test on z. So the 0.05 critical value is 1.96 and the 0.01 critical value is 2.576. Note that we have incorporated the continuity correction in computing the z-score because the values of U are integers, but we are approximating the distribution by the normal distribution, which is continuous (that's the $\frac{1}{2}$ that we subtracted from U). (See page 140 for a discussion of the continuity correction, if you missed it earlier.) We'll make the computation of a z-score for Example 1, even though the table would suffice:

$$\mu_U = \frac{(10)(10)}{2} = 50,$$

$$\sigma_U = \sqrt{\frac{(10)(10)(21)}{12}} = 13.23,$$

$$z = \frac{50 - 0 - \frac{1}{2}}{13.23} = \frac{49.5}{13.23} = 3.74.$$

Since z exceeds 2.576, we can reject the null hypothesis at the 0.01 level, which is what we already concluded using the U-table. If z had exceeded 1.96 but not 2.576, we would have rejected the null hypothesis at the 0.05 level of significance. If z had been less than 1.96, we would not have been able to reject the null hypothesis. ■

Example 2 Revisited

Again we rank the data:

Data:	0.7	1.0	1.1	1.5	2.1	2.2
Rank:	1	2	3	4	5	6
Sample:	*B*	*A*	*A*	*B*	*A*	*A*

Data:	2.3	2.7	3.5	3.6	4.0
Rank:	7	8	9	10	11
Sample:	*B*	*A*	*B*	*B*	*A*

We compute R_B because it looks easier: $R_B = 1 + 4 + 7 + 9 + 10 = 31$. Then

$$R_A = \frac{(11)(12)}{2} - 31 = 66 - 31 = 35 \text{ by equation (2)},$$

$$U_A = (6)(5) + \frac{1}{2}(6)(7) - R_A = 51 - 35 = 16,$$

$$U_B = (6)(5) + \frac{1}{2}(5)(6) - R_B = 45 - 31 = 14,$$

$$U = (\text{Smaller of 16 and 14}) = 14.$$

We look in Table 9 for critical values of U, with sample sizes 6 and 5. Notice that the table is organized in terms of n_1 and n_2 for the two sample sizes, and arranged so that $n_1 \leq n_2$. It makes no difference to us, or the statistic, which we call n_1 and which we call n_2. But to find the entry in the table, $n_1 \leq n_2$, so we *must* assign $n_1 = 5$ and $n_2 = 6$ or we won't find an entry in the table. We observe that the 0.05 critical value for U is 3, and the 0.01 critical value is 1. Since our value of $U = 14$ exceeds both of those, we cannot reject the null hypothesis. (*Note*: We compute two values for U, a smaller and a larger, and then always take the smaller, so that we can have a table of *small* values for U. If we wished to include an additional table of *large* values for U, we could compute just U_A or U_B and then see if the value we got was larger than the large value, or smaller than the small value. We chose to simplify the table instead.) ∎

IF THERE ARE TIES IN THE DATA

After ranking, two or more numbers may be the same:

Data:	. . . 3	3 . . .
Ranks:	. . . 4	5 . . .

In this case we take all the tied elements and average their ranks. We then assign that average number to all those tied elements and then proceed as above. Since $(4 + 5)/2 = 4.5$, we would have

Data: ... 3   3 ...
Ranks: ... 4.5 4.5 ....

EXERCISE 8   Use the Mann–Whitney, rank-sum test to test whether or not there is any difference in the effect of diet A or diet B on rats, referring to the data of Exercise 7. Remember to average the ranks of ties between diet A and diet B.

SEVERAL-POPULATION RANK-SUM: THE KRUSKAL–WALLIS TEST

We apply the same ideas as the Mann–Whitney U-test to the case of three or more populations. We require at least five numbers in each sample. We pool the data, as before, and compute the ranks after ordering the data in increasing order. Suppose R_i is the sum of the ranks of the elements from sample i, and n_i is the sample size for sample i. Then we compute the statistic

$$H = \frac{12}{n(n + 1)} \sum_{i=1}^{k} \frac{R_i^2}{n_i} - 3(n + 1),$$

where n is the total number of elements ($n = n_1 + n_2 + \cdots + n_k$). If the null hypothesis (that all the populations are identical) is true, then the distribution of H is closely approximated by a chi-square distribution, with $k - 1$ degrees of freedom, where k is the number of populations. So we compute H and see if the value exceeds the chi-squared 0.05 or 0.01 critical value for $(k - 1)$ d.f.

PROBLEM SET A   1. Given the following 12 observations, test the hypothesis that the median is greater than 6.00.

6.90   5.02   6.49   6.98   5.72   6.24   6.30   5.67
5.78   6.15   6.77   6.11

2. Given the following 5 observations, test the hypothesis that the median is greater than 5.

8   15   11   10   9

3. Given the following 22 observations, test the hypothesis that the median is greater than 10.8.

10.8	10.7	10.5	10.8	10.3	10.2	10.9	10.9
10.4	10.6	10.7	10.8	10.3	10.7	10.7	10.4
10.4	10.2	10.5	10.0	10.8	10.6		

4. Given the following 18 observations, test the hypothesis that the median is greater than 10.20.

10.13	10.56	10.40	10.42	10.92	10.20	10.69
10.13	10.29	10.81	10.91	10.55	10.01	10.92
10.05	10.04	10.09	10.28			

5. Given the following 10 observations, test the hypothesis that the median is greater than 11.0.

13.5	11.6	18.8	17.0	10.2	16.1	13.7	19.8
11.9	13.3						

6. Given the following 33 observations, test the hypothesis that the median is greater than 10.40.

10.80	10.23	10.10	10.90	10.46	10.36	10.56
10.19	10.80	10.87	10.65	10.27	10.73	10.85
10.70	10.50	10.21	10.22	10.45	10.95	10.07
10.30	10.52	10.04	10.09	10.54	10.13	10.45
10.35	10.54	10.05	10.22	10.77		

7. Given the following 9 observations, test the hypothesis that the median is greater than 13.

13	17	11	19	18	19	14	13	17

8. Given the following 24 observations, test the hypothesis that the median is greater than 22.0.

22.6	27.5	24.1	24.3	20.9	29.1	22.6	22.3
29.6	27.6	25.6	29.9	28.8	25.8	25.6	28.4
20.8	20.2	23.9	21.6	29.4	22.2	26.8	28.9

PROBLEM SET B

1. Given the following 12 observations, test the hypothesis that the median is equal to 6.00.

6.15	6.39	5.00	5.57	5.56	5.85	6.10	6.39
5.24	6.90	5.14	5.12				

2. Given the following 5 observations, test the hypothesis that the median is equal to 5.

11	12	15	9	7

3. Given the following 22 observations, test the hypothesis that the median is equal to 10.8.

10.6	10.4	10.5	10.5	10.6	10.6	10.8	10.0
10.6	10.4	10.6	10.0	10.1	10.7	10.9	10.8
10.8	10.6	10.5	10.9	10.9	10.7		

4. Given the following 18 observations, test the hypothesis that the median is equal to 10.20.

10.81	10.32	10.66	10.95	10.36	10.67	10.85
10.97	10.83	10.00	10.92	10.83	10.23	10.08
10.11	10.16	10.81	10.95			

5. Given the following 10 observations, test the hypothesis that the median is equal to 11.0.

19.2	17.5	11.7	16.3	18.1	15.7	15.7	11.8
11.5	17.9					

6. Given the following 33 observations, test the hypothesis that the median is equal to 10.40.

10.48	10.98	10.42	10.25	10.46	10.17	10.70
10.08	10.20	10.51	10.32	10.59	10.85	10.32
10.84	10.38	10.62	10.80	10.27	10.79	10.60
10.50	10.48	10.48	10.83	10.13	10.93	10.38
10.62	10.53	10.63	10.78	10.16		

7. Given the following 9 observations, test the hypothesis that the median is equal to 13.

19	13	17	13	16	19	17	17	15

8. Given the following 24 observations, test the hypothesis that the median is equal to 22.0.

24.3	25.5	21.0	21.9	23.6	23.9	28.6	22.2
25.4	23.3	29.2	27.3	22.2	28.8	27.4	20.9
28.8	29.3	27.3	29.4	21.6	28.8	27.6	24.6

PROBLEM SET C

1. Given the following 12 *paired* observations, test the hypothesis that the median of population A is less than the median of population B.

Sample A:	6.54	5.90	6.07	6.01	6.84	6.92	6.45
Sample B:	6.89	7.40	6.54	6.29	8.08	7.32	7.50
Sample A:	6.08	5.14	5.06	6.40	6.36		
Sample B:	7.72	5.40	6.43	7.98	6.61		

2. Given the following 25 *paired* observations, test the hypothesis that the median of population A is less than the median of population B.

Sample A:	10	14	7	14	9	7	14	13	8
Sample B:	10	14	8	15	10	8	15	13	8

Sample A:	14	10	10	14	8	12	15	8	14
Sample B:	15	10	11	14	7	11	16	8	15

Sample A:	15	9	9	14	6	13	12
Sample B:	15	10	10	14	7	13	12

3. Given the following 18 *paired* observations, test the hypothesis that the median of population A is less than the median of population B.

Sample A:	10.37	10.31	10.49	10.65	10.08	10.28
Sample B:	10.66	11.28	11.40	10.47	10.77	11.06

Sample A:	10.84	10.12	10.81	10.26	10.29	10.92
Sample B:	12.39	11.45	12.14	10.44	10.71	12.00

Sample A:	10.95	10.74	10.23	10.30	10.98	10.89
Sample B:	11.86	12.02	10.54	10.71	12.61	10.82

4. Given the following 10 *paired* observations, test the hypothesis that the median of population A is less than the median of population B.

Sample A:	13.4	15.1	19.2	11.2	11.4	12.2	19.0
Sample B:	13.4	16.0	19.4	12.0	13.0	13.8	19.9

Sample A:	15.2	14.8	10.6
Sample B:	15.5	15.6	10.3

5. Given the following 33 *paired* observations, test the hypothesis that the median of population A is less than the median of population B.

Sample A:	10.63	10.39	10.03	10.12	10.88	10.71
Sample B:	11.29	10.12	10.66	11.77	12.52	10.69

Sample A:	10.81	10.45	10.01	10.71	10.63	10.52
Sample B:	10.85	10.59	10.78	11.59	11.66	11.78

Sample A:	10.44	10.29	10.67	10.80	10.20	10.54
Sample B:	10.87	10.27	12.32	12.14	11.77	12.16

Sample A:	10.91	10.17	10.37	10.30	10.67	10.50
Sample B:	11.88	10.17	11.01	10.23	10.75	11.61

Sample A:	10.62	10.57	10.10	10.08	10.70	10.08
Sample B:	11.24	10.38	11.21	9.92	10.75	10.01

Sample A: 10.66   10.17   10.94
Sample B: 10.89   11.82   10.90

6. Given the following 19 *paired* observations, test the hypothesis that the median of population A is less than the median of population B.

Sample A:	17	16	14	17	11	17	17	10	12	13
Sample B:	16	16	15	17	11	18	16	10	12	14

Sample A:	19	19	14	12	12	11	15	10	11
Sample B:	20	19	15	11	12	11	16	10	12

7. Given the following 24 *paired* observations, test the hypothesis that the median of population A is less than the median of population B.

Sample A:	22.3	29.0	27.4	20.7	25.8	22.5	20.7
Sample B:	23.1	29.7	29.0	20.4	27.3	23.9	21.2

Sample A:	20.6	25.3	28.9	21.4	29.3	20.5	23.9
Sample B:	21.9	26.2	28.9	21.2	30.9	20.9	24.0

Sample A:	28.6	20.3	21.7	29.8	25.0	26.6	24.7
Sample B:	29.7	20.1	22.4	31.1	24.8	27.6	25.9

Sample A:	21.2	21.4	25.4
Sample B:	21.0	21.9	25.2

8. Given the following 22 *paired* observations, test the hypothesis that the median of population A is less than the median of population B.

Sample A:	10.6	10.9	10.1	10.3	10.9	10.8	10.8
Sample B:	11.4	10.6	11.6	11.3	11.9	10.7	11.4

Sample A:	10.3	10.8	10.9	10.7	10.4	10.0	10.3
Sample B:	11.0	11.5	11.8	12.1	11.7	11.4	10.5

Sample A:	10.3	10.4	10.6	10.2	10.1	10.5	10.2
Sample B:	10.6	10.9	10.7	11.2	10.3	10.8	10.6

Sample A:	10.5
Sample B:	11.7

PROBLEM SET D 1. Given the following 12 *paired* observations, test the hypothesis that the median of population A is equal to the median of population B.

Sample A:	5.97	5.67	6.38	6.75	6.94	6.73	5.10
Sample B:	6.89	5.47	6.26	7.91	7.73	8.32	5.54

Sample A:	5.90	5.00	6.74	5.99	6.61
Sample B:	6.16	6.53	6.66	7.14	7.19

2. Given the following 25 *paired* observations, test the hypothesis that the median of population *A* is equal to the median of population *B*.

Sample *A*:	9	6	7	9	12	12	10	9	6
Sample *B*:	10	7	7	9	12	12	10	9	6

Sample *A*:	15	10	8	13	6	8	12	8	13
Sample *B*:	15	9	8	13	6	8	12	7	13

Sample *A*:	6	7	6	13	15	8	7
Sample *B*:	7	7	6	13	15	8	6

3. Given the following 18 *paired* observations, test the hypothesis that the median of population *A* is equal to the median of population *B*.

Sample *A*:	10.62	10.69	10.93	10.33	10.64	10.51
Sample *B*:	12.06	10.69	11.29	10.94	11.58	10.26

Sample *A*:	10.87	10.11	10.11	10.13	10.08	10.75
Sample *B*:	11.34	11.40	10.27	10.44	9.85	11.68

Sample *A*:	10.91	10.54	10.45	10.03	10.01	10.43
Sample *B*:	11.69	11.32	10.51	9.91	10.02	10.47

4. Given the following 10 *paired* observations, test the hypothesis that the median of population *A* is equal to the median of population *B*.

Sample *A*:	11.7	16.8	19.6	16.9	12.4	18.8	13.7
Sample *B*:	12.9	17.5	21.1	16.6	12.4	20.3	14.4

Sample *A*:	19.2	10.3	17.6
Sample *B*:	20.4	11.8	17.9

5. Given the following 33 *paired* observations, test the hypothesis that the median of population *A* is equal to the median of population *B*.

Sample *A*:	10.44	10.91	10.71	10.91	10.46	10.16
Sample *B*:	11.83	11.84	12.39	11.66	11.21	10.82

Sample *A*:	10.77	10.87	10.28	10.37	10.56	10.63
Sample *B*:	11.10	11.32	11.01	11.62	11.09	11.12

Sample *A*:	10.83	10.96	10.74	10.13	10.62	10.53
Sample *B*:	11.55	12.22	12.08	10.35	11.27	10.56

Sample *A*:	10.79	10.78	10.25	10.40	10.53	10.92
Sample *B*:	11.60	11.02	10.23	11.08	11.57	12.35

Sample *A*:	10.88	10.08	10.61	10.37	10.76	10.79
Sample *B*:	11.57	11.21	11.77	10.33	11.07	11.83

Sample *A*:	10.99	10.87	10.71
Sample *B*:	11.33	11.16	11.60

6. Given the following 19 *paired* observations, test the hypothesis that the median of population *A* is equal to the median of population *B*.

Sample *A*:	19	14	16	16	16	12	18	10	13
Sample *B*:	19	14	16	16	16	12	19	9	13

Sample *A*:	11	11	14	19	15	10	10	16	15
Sample *B*:	12	10	13	20	14	10	10	16	15

Sample *A*:	15
Sample *B*:	15

7. Given the following 24 *paired* observations, test the hypothesis that the median of population *A* is equal to the median of population *B*.

Sample *A*:	29.5	25.6	26.2	27.4	20.4	26.8	29.8
Sample *B*:	30.6	25.7	26.6	27.9	20.7	27.9	29.9

Sample *A*:	29.6	21.8	29.1	22.4	26.3	28.0	22.8
Sample *B*:	31.2	23.4	30.3	23.7	26.6	29.5	23.9

Sample *A*:	28.4	25.7	24.3	25.4	25.2	25.7	21.7
Sample *B*:	28.7	26.1	24.0	26.6	25.0	26.0	22.3

Sample *A*:	26.5	28.8	28.8
Sample *B*:	27.4	29.9	30.3

8. Given the following 22 *paired* observations, test the hypothesis that the median of population *A* is equal to the median of population *B*.

Sample *A*:	10.2	10.2	10.8	10.7	10.3	10.2	10.0
Sample *B*:	10.2	11.5	10.7	10.6	10.8	10.3	10.7

Sample *A*:	10.6	10.3	10.1	10.7	10.3	10.7	10.7
Sample *B*:	11.1	10.3	11.1	10.7	11.0	11.6	11.0

Sample *A*:	10.6	10.9	10.8	10.5	10.0	10.2	10.9
Sample *B*:	12.2	12.1	12.1	10.9	11.0	11.2	10.7

Sample *A*:	10.1
Sample *B*:	10.2

CHAPTER 8 QUIZ   1. (True-False)   To perform analysis of variance, you need to assume normality of populations. To use the one- or two-sample sign test, you must make the same assumption.

2. The sign test is a test about the _____ .

3. The mean and the median of a population are the same if the distribution of the population is _____ .

4. With paired data, you can make a _____ test.

5. With unpaired data, you can pair the data by use of _____ .

6. The Mann–Whitney test tests the null hypothesis that the two populations

_____ .

7. The Mann–Whitney U-statistic measures the degree of _____ in the ranking of the two samples.

ANSWERS TO EXERCISES

1. Comparing our eight travel times with 60 minutes, we observe

59	48	46	55	61	47	48	58
−	−	−	−	+	−	−	−

which is 7 minuses and 1 plus. Since the hypothesis is "more than," we make a *one-tailed test*.

   a. We look in Table 7a and see that at the 0.01 level, we need 0 or 8 pluses. We cannot reject at the 0.01 level of significance.

   b. At the 0.05 level, we need 1 or fewer or 7 or more pluses in order to reject $\mu = 60$ (one-tailed). We reject $\mu = 60$ at the 0.05 level and conclude $\mu < 60$.

2. We perform the same analysis as in Exercise 1, but now we make a two-tailed test (because "equals" versus "not equal" leads to a two-sided alternative). We observe in Table 7a that to reject the hypothesis at the 0.05 level, we would need to observe either 0 or 8 pluses. We cannot therefore reject the hypothesis at the 0.05 level or the 0.01 level.

3. Comparing the 18 breaking strengths with 55, we observe

58	47	53	56	58	60	56	49	54
57	51	54	56	50	56	55	59	57
+	−	−	+	+	+	+	−	−
+	−	−	+	−	+		+	+

We have 10 pluses, 7 minuses, and 1 tie. Since $n > 15$ (actually, 17), we use Table 7b. $k =$ larger number of pluses or minuses $= 10$.

$$z = \frac{2(10) - 1 - 17}{\sqrt{17}} = 0.49$$

This is a two-tailed test ("equals" versus "not equal"), and the value of z does not exceed 1.96 or 2.576, so it is not significant at either the 0.05 level or the 0.01 level. We cannot reject $\mu = 55$. The data do support the claim.

4. The analysis is the same as in Exercise 3, but leads to a one-tailed test. We compare $z = 0.49$ with 1.645 and 2.33 and see that it is not significant. We cannot reject $\mu \geqslant 55$.

5. From Table 7b,

$$z = \frac{2(57) - 1 - 100}{\sqrt{100}} = 1.3;$$

not significant at the 0.05 level.

6. This is a two-sample sign test—paired data (paired because we have two scores from each of the 15 golfers). We observe 12 pluses, 1 minus, and 2 ties. As the hypothesis is formulated, it is a one-tailed test ("*lower* golf scores"). We look in Table 7a and observe that for $n = 13$, one-tailed, the 0.01 level critical value is 1 or fewer pluses or 12 or more pluses. Since our data meet this critical value, we can reject "equal" golf scores and conclude that the data support the hypothesis of lower golf scores with the new ball, at the 0.01 level of significance.

7. The answer you get will depend on which random numbers you happen to pick.

   Step 1.   Pick 10 random two-digit numbers from Table 8, and associate one of them to each of the 10 weight gains from diet B (the *larger* of the two samples). If any of the two-digit numbers are the same, attach a third digit to each to see which should come first in the sort. If any of the three-digit numbers are the same, attach a fourth digit, and so on. I used the first 20 digits of the first row of Table 8 for the 10 two-digit numbers:

Random numbers:	17	74	80	49	63
	↕	↕	↕	↕	↕
Diet B:	21	25	27	29	25

Random numbers:	54	85	98(4)	86	98(1)
	↕	↕	↕	↕	↕
Diet B:	24	27	29	28	26

   Notice that the eighth and tenth random numbers are both 98. We take the *next two* digits from the table, 4 and 1. Thus the first 98 becomes 98(4) and the second becomes 98(1), allowing us to complete the sort.

   Step 2.   Now sort the data of diet B in increasing order according to the random numbers, and then match the first 7 of these data with the 7 weight gains of diet A *in their original order*. Then proceed as with the paired data.

Random numbers sorted:	17	49	54	63	74	80	85	86	98(1)	98(4)
	↕	↕	↕	↕	↕	↕	↕	↕	↕	↕
Diet B:	21	29	24	25	25	27	27	28	26	29
Diet A:	15	21	17	19	23	24	19			
	+	+	+	+	+	+	+			

   We observe 7 pluses, $n = 7$. This is a two-tailed (". . . no difference . . .") test. Checking Table 7, we see that we can reject at the 0.05 level of significance.

8. Pool the two sets of weight gains and then sort in increasing order:

	15	17	19	19	21	21	23	24	24
	A	A	A — A		A — B		A	A — B	
Rank:	1	2	3.5	3.5	5.5	5.5	7	8.5	8.5

25	25	26	27	27	28	29	29

	B	$-$	B	B	B	$-$	B	B	B	$-$	B
Rank:	10.5		10.5	12	13.5		13.5	15	16.5		16.5

$$R_A = 31; \quad n = 7 + 10 = 17$$

$$R_B = \frac{17(18)}{2} - 31 = 122$$

$$U_A = 10(7) + \frac{7(8)}{2} - 31 = 67$$

$$U_B = 10(7) + \frac{10(11)}{2} - 122 = 3$$

$$U = \text{Minimum of } (67 \text{ and } 3) = 3$$

In Table 9, the critical value for samples sizes (7, 10) is 14 at the 0.05 level and 9 at the 0.01 level. Since $U = 3$ is smaller than both of these, we can reject the null hypothesis that the two populations are identical at the 0.01 level. We conclude that the diets have different effects.

ANSWERS TO CHAPTER 8 QUIZ

1. False—that's why it's a *nonparametric* test.

2. median

3. symmetric

4. sign

5. random numbers

6. are identical

7. mixing

Confidence Intervals for Standard Deviation and Variance

Terms we'll be learning about in this chapter

option writer

striking price

volatility

9-1 INTRODUCTION

In the stock market there is a financial instrument called an option, which you can buy or sell like common stock. The option allows its owner, if he or she so chooses, to buy a certain stock for a specified price, called the **striking price**, from the seller of the option. The seller of the option is called the **option writer**. The option expires after a certain amount of time, normally from 3 to 9 months after it is written. Once the option expires it has no value.

Let's suppose I purchase an option to buy 100 shares of IBM stock at a striking price of $60 per share, with the option expiring next December 15. Suppose that the current selling price of IBM is $58 per share. And suppose that I pay the option writer $400 for this option. Why would I do such a thing? There would be no point in my exercising my option to buy IBM from the option writer at $60 per share if I can buy it in the marketplace for $58 per share. But what I am hoping for is that the price of IBM will be $70, or even $80, per share before December 15. If the price were $70 per share, then my option would be worth $10 per share, or $1000 (as the option allows me to buy the stock from the option writer at $60 per share and then turn around and sell it in the marketplace for $70 per share, or a profit of $10 per share times 100 shares). If the price went to $80 per share, my option would be worth $2000. Such great increases in the worth of the option are why people are willing to pay substantial sums of money to buy options that are worth nothing if exercised right now. These people hope that the options will *become* valuable. And from the option seller's point of view, he or she receives the cash *now* for this option and hopes that the stock doesn't increase too much over the next several months. In fact, if the stock price never moves from $58 per share, the seller will pocket my $400 and have to pay out nothing in return. (But if the stock runs up to $70 per share, I will exercise my option and it will cost him $1000.)

What I am willing to pay for such an option, and what an option writer is willing to sell such an option for, depends on how likely each of us thinks it is for the stock to increase substantially in price over the next several months. (Actually, there are options that allow you to "win," or profit, if the stock goes down as well.) The likelihood of a stock's moving up or down a substantial amount over a period of time depends on what is called its volatility. The **volatility** of a stock is a measure of how rapidly moving the price of the stock has been or will be. The more likely it is that the stock is going to increase greatly in the next few months, the more the buyer of the option is willing to pay, and the greater the risk (potential loss) there is to the seller. In my case, from the seller's point of view, the best thing that could happen is that IBM stays between $58 and $60 per share from now through December 15— that is, that IBM would have *little volatility*. So you can see that before writing an option on a particular stock, it's very important to have a good measure of the volatility of the stock.

One of the best measures of volatility is the variance of stock prices. If you think about it for a moment, this will come as no surprise. Volatility measures movement, or variation, in stock price; variance also measures variation. For this reason, being able to estimate the variance is an important statistical application in the stock market. As you already know, the scientifically correct method of estimation is to form confidence intervals. This is the topic of this chapter: confidence intervals for variance and standard deviation.

9-2 THE ESTIMATION OF σ^2

If you recall, when we estimated the population mean, we found we had to know the population variance in order to know the variance of the DSM. And when the sample size was 30 or greater, we simply took the sample variance as the population variance, even though we knew that this was not exactly correct. When the sample size was less than 30, we corrected for this substitution by use of the t-distribution. But in each case, we were using s^2, the sample variance, as an estimate of σ^2, the population variance.

Just how good an estimate is s^2 for σ^2? The answer depends first on the sample size. The larger the sample size, the better estimate s^2 is for σ^2. Let's fix the sample size for the moment at n. Then s^2, being a statistic, has a distribution, just as the DSM was the distribution for sample means (of some fixed size). Now the distribution of s^2 cannot have the same shape as the DSM. For one thing, s^2 is a sum of squares of numbers and so can never be negative. The distribution of s^2 will be different for each different sample size, for each different population, and for each different population variance.

However, things simplify considerably, since most populations whose distributions are near normal in shape give rise to the same shape distribution for s^2, given one fixed sample size. That distribution is a chi-square distribution, $(n - 1)$ d.f., and looks like the curve in Figure 9-1. Notice that the "hump" in the distribution occurs directly above $(n - 1)\sigma^2$, which means that the "most likely" value for s^2 is $(n - 1)\sigma^2$.

Actually, there isn't any reason to have a different chi-square distribution for

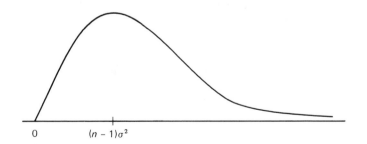

0    $(n - 1)\sigma^2$

Distribution of s^2, sample size = n
chi-square distribution, $(n - 1)$ d.f.

Figure 9-1

each possible value of σ^2. We can "normalize" these distributions, just as we did with the normal distribution. Instead of considering the distribution of s^2, we consider the distribution of

$$\frac{(n-1)s^2}{\sigma^2},$$

which has exactly the same shape, but now has its hump at $n-1$ (see Figure 9-2).

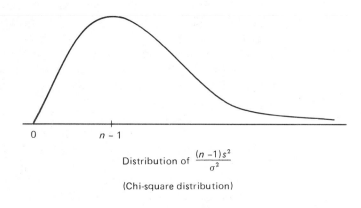

Distribution of $\dfrac{(n-1)s^2}{\sigma^2}$

(Chi-square distribution)

Figure 9-2

Now, to form a 90% confidence interval for σ^2 and σ, we use the chi-square distribution as follows. Suppose the sample size, n, is 10. By using Table 4, a table of critical values of chi-square (at the back of the book), we find the point on the number line where 5% of the distribution lies to the right (that is, 5% in the right-hand tail) and the point where 95% lies to the right (5% in the left-hand tail). These are called the **5%** and **95% critical values** of the distribution, denoted $\chi^2_{0.05}$ and $\chi^2_{0.95}$. (See Figure 9-3.) Table 4 tells us that the 95% and 5% critical values of chi-square, 9 d.f. (degrees of freedom, one less than the sample size, which we sup-

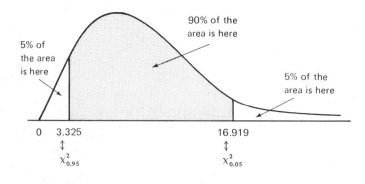

Figure 9-3

posed to be 10), are 3.325 and 16.919. See Figure 9-3 again. This information is recorded on the ninth line of Table 4, as follows:

n	d.f.	$\chi^2_{0.95}$	$\chi^2_{0.05}$
10	9	3.325	16.919

↑   ↑
95% critical   5% critical
value   value

These two values are the left- and right-hand endpoints of a 90% confidence interval for

$$\frac{(n-1)s^2}{\sigma^2}$$

(see Figures 9-4 and 9-5, page 276).

$$3.325 \le \overset{\overset{\displaystyle (n-1)}{\downarrow}}{\frac{9s^2}{\sigma^2}} \le 16.919 \tag{1}$$

Taking reciprocals,

$$\frac{1}{3.325} \ge \frac{\sigma^2}{9s^2} \ge \frac{1}{16.919}. \tag{2}$$

Multiplying by $9s^2$,

$$\frac{9s^2}{3.325} \ge \sigma^2 \ge \frac{9s^2}{16.919}. \tag{3}$$

Now reversing the inequalities,

$$\frac{9s^2}{16.919} \le \sigma^2 \le \frac{9s^2}{3.325}. \tag{4}$$

This is a 90% confidence interval for σ^2. So if the sample variance s^2 from the sample of size 10 is $s^2 = 1000$, for example, then we substitute 1000 for s^2 in (4) and obtain

$$\frac{9000}{16.919} \le \sigma^2 \le \frac{9000}{3.325}, \tag{5}$$

or

$$531.94 \le \sigma^2 \le 2706.77. \tag{6}$$

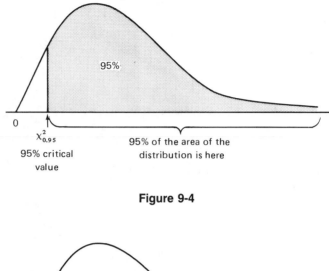

95%

0

$\chi^2_{0.95}$

95% critical
value

95% of the area of the
distribution is here

Figure 9-4

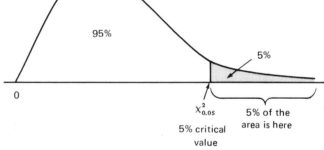

95%

5%

0

$\chi^2_{0.05}$

5% critical
value

5% of the
area is here

Figure 9-5

Taking square roots,

$$23.06 \leq \sigma \leq 52.02, \qquad (7)$$

which gives our 90% confidence interval for σ, and (6) the same for σ^2.

To review, we used the numbers 3.325 and 16.919, which we said were the 95% and 5% critical values for chi-square ($n = 10$), respectively. This means that when a number is drawn (at random) from the distribution of chi-square, 95% of the time that number will *exceed* the number 3.325 (that is, be to the right of the number 3.325 in Figure 9-3). Similarly, when a number is drawn at random, exactly 5% of the time it will exceed the value 16.919. That's why these are called critical values of the distribution. It's not necessary to display the entire distribution of chi-square. Its critical values will be sufficient. A table of chi-square will give the critical values for 0.995, 0.99, 0.975, 0.95, 0.05, 0.025, 0.01, and 0.001 at each sample size $n = 2$ to $n = 30$.

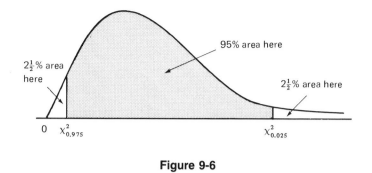

Figure 9-6

To form a 95% confidence interval instead of a 90% confidence interval, we would seek the 0.975 and 0.025 critical values for chi-square, 9 d.f. (These are the left- and right-hand endpoints, respectively, of a 95% confidence interval. See Figure 9-6.) From (1) through (5), we see that to form the 95% confidence interval we would write

$$\chi^2_{0.975} \leqslant \frac{9s^2}{\sigma^2} \leqslant \chi^2_{0.025}, \qquad (8)$$

leading to

$$\frac{9(1000)}{\chi^2_{0.025}} \leqslant \sigma^2 \leqslant \frac{9(1000)}{\chi^2_{0.975}}. \qquad (9)$$

We refer again to Table 4, the table of critical values of chi-square, 9 d.f.:

n	d.f.	$\chi^2_{0.995}$	$\chi^2_{0.99}$	$\chi^2_{0.975}$	$\chi^2_{0.95}$	$\chi^2_{0.05}$	$\chi^2_{0.025}$	$\chi^2_{0.01}$	$\chi^2_{0.005}$
10	9	1.735	2.088	2.700	3.325	16.919	19.023	21.666	23.589

                ↑   ↑

       Left end   Right end

(In this table, the number under $\chi^2_{0.975}$ is the value on the number line such that 97.5% of the distribution lies to the right, and hence 2½% of the distribution lies to the left, and so forth.)

Thus, we get from (9),

$$\frac{9000}{19.023} \leqslant \sigma^2 \leqslant \frac{9000}{2.700}, \qquad (10)$$

or, after taking square roots,

$$21.75 \leqslant \sigma \leqslant 57.74. \qquad (11)$$

We can repeat this for any sample size and confidence level. Remember that the 9 in formula (8) is really $n - 1$, one less than the sample size. I showed you how to form a 90% confidence interval in (4), which involved the 95% and 5% critical values of χ^2. Then I showed you how to form a 95% confidence interval in (9), which involved the 97.5% and 2.5% critical values of χ^2. In general, if we want a $P\%$ confidence interval, where P is a number between 0 and 100, then we set

$$p = \frac{P}{100}$$

(for example, $P = 90$ and $p = 0.90$, or $P = 95$ and $p = 0.95$, in our two cases above). Then we use the

$$\left(\frac{100 + P}{2} \right)\% \quad \text{and} \quad \left(\frac{100 - P}{2} \right)\% \quad \text{critical values}$$

$$\chi^2_{(1 + p)/2} \quad \text{and} \quad \chi^2_{(1 - p)/2}; \quad n - 1 \text{ d.f.;}$$

for a $P\%$ confidence interval:

$$\frac{(n - 1)s^2}{\chi^2_{(1 - p)/2}} \leqslant \sigma^2 \leqslant \frac{(n - 1)s^2}{\chi^2_{(1 + p)/2}}. \tag{12}$$

9-3 LARGE-SAMPLE ESTIMATION OF σ

If $n > 30$, we do not use the table of critical values of chi-square. Instead, we use the fact that s is approximately normally distributed, with mean $= \sigma$ and standard deviation $= \sigma/\sqrt{2n}$. Thus, if n is 50, a 95% confidence interval for σ is

$$s \pm 1.96 \frac{s}{\sqrt{2(50)}}.$$

In general, a 95% confidence interval would be

$$s \pm 1.96 \frac{s}{\sqrt{2n}}. \tag{13}$$

EXERCISE 1   A sample of size 5 is drawn from a normal population. The sample variance is $s^2 = 100$. Compute 90%, 95%, and 99% confidence intervals for σ. Use (12) and take square roots.

EXERCISE 2 A sample of size 26 is drawn from a normal population. The sample variance is $s^2 = 60$. Compute 90%, 95%, and 99% confidence intervals for σ.

_____   _____   _____

EXERCISE 3 A sample of size 80 is drawn from a normal population. The sample variance is $s^2 = 40$. Compute 90%, 95%, and 99% confidence intervals for σ. [Careful—this is a "large sample." Use (13).]

_____   _____   _____

PROBLEM SET A
1. A sample of size 15 is drawn from a normal population. The sample variance is 400. Compute 90%, 95%, and 99% confidence intervals for σ.

2. A sample of size 25 is drawn from a normal population. The sample variance is 1000. Compute 90%, 95%, and 99% confidence intervals for σ.

3. A sample of size 60 is drawn from a normal population. The sample variance is 1000. Compute 90%, 95%, and 99% confidence intervals for σ.

4. A sample of size 240 is drawn from a normal population. The sample variance is 100. Compute 90%, 95%, and 99% confidence intervals for σ.

5. A sample of size 12 is drawn from a normal population. The sample variance is 1. Compute 90%, 95%, and 99% confidence intervals for σ.

6. A sample of size 20 is drawn from a normal population. The sample variance is 400. Compute 90%, 95%, and 99% confidence intervals for σ.

CHAPTER 9 QUIZ
1. s^2 is an _____ for σ^2.

2. The distribution of s^2, actually of $\dfrac{(n-1)s^2}{\sigma^2}$, is the _____ distribution.

3. The value such that 5% of the chi-square distribution lies to the right of that value is called the _____ .

4. When we have a sample of size 30 or more, we do not use the _____ _____ to form a confidence interval estimate for the population variance. Instead, we use the fact that s is approximately _____ .

ANSWERS TO EXERCISES
1. 90%: $n = 5$, $n - 1 = 4$ d.f.;

$P = 90$, $p = 0.90$;

$\dfrac{1-p}{2} = 0.05$, $\dfrac{1+p}{2} = 0.95$;

at 4 d.f., $\chi^2_{0.95} = 0.711$;   at 4 d.f., $\chi^2_{0.05} = 9.488$;

$$\frac{4(100)}{9.488} \leq \sigma^2 \leq \frac{4(100)}{0.711} ;\quad 6.49 \leq \sigma \leq 23.72$$

95%: $P = 95$, $p = 0.95$;

$$\frac{1-p}{2} = 0.025, \quad \frac{1+p}{2} = 0.975;$$

at 4 d.f., $\chi^2_{0.975} = 0.484$;   at 4 d.f., $\chi^2_{0.025} = 11.143$;

$5.99 \leq \sigma \leq 28.75$

99%: $P = 99$, $p = 0.99$;

$$\frac{1-p}{2} = 0.005, \quad \frac{1+p}{2} = 0.995;$$

at 4 d.f., $\chi^2_{0.995} = 0.207$;   at 4 d.f., $\chi^2_{0.005} = 14.860$;

$5.19 \leq \sigma \leq 43.96$

2. $n = 26$, $n - 1 = 25$ d.f.

90%:   At 25 d.f., $\chi^2_{0.95} = 14.611$;   at 25 d.f., $\chi^2_{0.05} = 37.652$;

$$\frac{25(60)}{37.652} \leq \sigma^2 \leq \frac{25(60)}{14.611} ;\quad 6.31 \leq \sigma \leq 10.13$$

95%:   At 25 d.f., $\chi^2_{0.975} = 13.120$;   at 25 d.f., $\chi^2_{0.025} = 40.646$;

$6.07 \leq \sigma \leq 10.69$

99%:   At 25 d.f., $\chi^2_{0.995} = 10.520$;   at 25 d.f., $\chi^2_{0.005} = 46.928$;

$5.65 \leq \sigma \leq 11.94$

3. 90%: $s \pm 1.645 \dfrac{s}{\sqrt{2n}} = 6.32 \pm 1.645 \dfrac{6.32}{\sqrt{160}}$

$\qquad\qquad\qquad = 6.32 \pm 0.82$

95%: $s \pm 1.96 \dfrac{s}{\sqrt{2n}} = 6.32 \pm 1.96 \dfrac{6.32}{\sqrt{160}}$

$\qquad\qquad\qquad = 6.32 \pm 0.98$

99%: $s \pm 2.576 \dfrac{s}{\sqrt{2n}} = 6.32 \pm 2.576 \dfrac{6.32}{\sqrt{160}}$

$\qquad\qquad\qquad = 6.32 \pm 1.28$

ANSWERS TO CHAPTER 9 QUIZ

1. estimator

2. chi-square

3. 5% critical value of chi-square

4. chi-square distribution; normally distributed

Quick Word Guide

alternative hypothesis The negative of the null hypothesis.

ANOVA Analysis of variance.

bar chart See *histogram*.

bell-shaped curve See *normal distribution*.

biased estimator An estimator that does not approximate the parameter.

bimodal Having two modes.

binomial distribution The "histogram" of possible outcomes from a binomial process of some specified number of trials, with a fixed probability of "success" on each.

binomial population A population consisting entirely of 0s and 1s.

binomial process A sequence of trials performed, each of which has the same likelihood of "success" (1) or "failure" (0), and for which the outcome of one trial in no way affects the outcome of any other trial.

binomial trials process See *binomial process*.

bivariate data Data that occur in ordered pairs, such as (height, weight); two-variable data.

cells The categories in a contingency table, or the categories in a two-way analysis of variance; the "entries" or items in a matrix.

central limit theorem The mathematical theorem stating that for any population with finite variance, if the sample size is large enough, then the distribution of sample means will be approximately normal.

chi-square distribution The distribution of the chi-square statistic.

chi-square statistic The statistic that measures the discrepancy between the observed values and the expected values in a contingency table.

class   A collection of numbers, usually all close together, grouped for purposes of constructing a histogram of a population; for example, the class of tire mileages between 37,000 and 38,000.

coin toss   The result of an imaginary experiment of tossing a fair coin and allowing it to fall to rest so that the outcome of the coin facing "heads" or "tails" is due to chance and cannot be predicted in advance.

confidence interval   An interval of numbers that locates a parameter along the number line, together with a confidence level.

confidence level   The percentage of time that a statement about the location of some parameter will be correct.

contingency table   The rectangular array resulting from the partitioning of two variables, such as political preference and age, into classes.

continuity correction   The process of adding or subtracting ½ from an integer value before computing a z-score, to account for the discreteness of the histogram and its associated classes.

continuous histogram   A curve, rather than a bar chart.

correlation   The degree of linear relationship between bivariate data or a bivariate population.

correlation coefficient   A statistic or parameter that measures the degree of correlation.

count data   Integer values that count the number of occurrences in the data of each category, usually in a contingency table.

critical value   A value that specifies the percentile (0.05 critical value = 95th percentile; 0.01 critical value = 99th percentile; and so on) of the distribution of some given statistic.

crop yields   The numbers describing the amount of a certain crop measured, for example, in tons per acre.

cure rate   The percentage of individuals who are cured by use of a certain drug.

decile   One of the division points between 10 equal-sized pieces of the population when the population is arranged in numerical order. The eighth decile is the number such that 8/10 of the population is smaller and 2/10 is larger.

degrees of freedom   A technical term measuring the number of freely, or independently, varying data. Applies to the chi-square distribution, the F-distribution, and the t-distribution.

dependent variable   The variable whose value is determined when the value of the (one or more) independent variable(s) is (are) known. Usually the variable of the vertical axis, or y-axis.

descriptive statistics   The study of descriptions of populations, either numerically or pictorially.

destructive testing   The process of obtaining numbers from objects by sampling in which it is necessary to "use up," or destroy, the objects; for example, to determine the lifetime of a light bulb, we must wear it out.

d.f.   See *degrees of freedom*.

distribution   A histogram of a statistic.

distribution of sample means   The histogram that records where the sample means of a certain fixed sample size fall, and how often, assuming that the sampling procedure is repeated thousands of times.

DSM   Distribution of sample means.

error   In the formation of confidence intervals, the percentage of the time that we incorrectly announce the location of the parameter. In hypothesis testing, false rejection or acceptance of a hypothesis (see *type 1* and *type 2 error*). In ANOVA, a source of variation in the data from *within* populations.

estimator   A statistic used to obtain information about a parameter.

event   Some specified collection of possible outcomes of some experiment, usually determined by some numerical values (for example, the event "57 or more successes out of 100 trials").

expectation   The mean of the distribution of a statistic.

expected numbers   The numbers we would observe, on the average, if the null hypothesis (of no relationship between the variables of the contingency table) is true.

factorial   The product of the numbers 1, 2, 3, and so on, up to the given number. The given number must be a positive integer or 0. By definition, 0 factorial is 1.

factorial key   A key on a calculator that will calculate factorial.

fair coin   A coin for which the likelihood of "heads" and "tails" is the same.

fractile   One of the division points between a specified number of equal-sized pieces of the population when the population is arranged in numerical order. For example, percentile, quartile, decile.

***F*-ratio**   A statistic used to test whether or not several populations all have the same mean. See analysis of variance, Chapter 6.

frequency   The number of occurrences. See also *relative frequency*.

***F*-test**   Test of the hypothesis that there is no difference in the population means, using the *F*-ratio as the test statistic.

function   A rule that relates the values of one variable, called the *independent variable*, to the values of another variable, called the *dependent variable*.

Gaussian elimination   The method of systematically transforming a system of linear equations to a form where they can be solved.

histogram   A diagram that indicates the relative frequency or percentage of different groupings of numbers, called classes, in the population. Also called a bar chart. See pages 3 and 4 for examples.

hypothesis   A statement about some population parameter that is to be tested for its correctness.

hypothesis testing   The procedure of forming a hypothesis and testing it by the use of some statistic.

independent trials process   See *binomial process*.

independent variable   The variable over which we have control, normally the variable of the horizontal axis, or *x*-axis.

inferential statistics   The science of making an inference about the whole population from information about only a part of the population, a sample.

interval   See *class*.

least squares   The criterion that measures which straight line "best fits" the given data; refers to the minimum possible value of the sum of the squared residuals of predicted *y*-values from observed *y*-values.

level of significance   See *significance level*.

linear function   A function whose graph is a straight line, or in the case of a function of several variables, the sum of constant multiples of the variables; for example: $y = 2x + 3v + 7u + 9$. Here y is a linear function of the variables x, v, and u.

location   The position along the number line, or *x*-axis.

Mann–Whitney   A statistical test based on rank-sum.

marginal return   The amount of increase we observe in the dependent variable when we increase the independent variable one unit.

maximum   The largest number in a population or sample.

mean   The average of a collection of numbers, obtained by adding all the numbers and then dividing by the total number of numbers.

mean square error   A statistic in analysis of variance; it measures "within-samples" variance.

median   The number such that half the population (or sample) is larger and half is smaller; also, the 50th percentile.

method of least squares   See *least squares*.

minimum   The smallest number in a population or sample.

mode   The most frequently occurring number in a population or sample. The mode of a population is a parameter.

MSE   Mean square error.

negative correlation   A relationship in which one variable decreases when the other variable increases. The marginal return is negative.

normal approximation   Approximation of a given distribution, such as the DSM or binomial distribution, by a normal distribution.

normal curve   The normal distribution.

normal distribution   The pattern that was discovered from repeated measurements and that has the shape of the bell-curve. See page 63.

normal equations   The linear equations we must solve to obtain the regression coefficients.

normal table   The table of values of the bell-shaped curve. See Tables 1 and 2 at the back of the book.

null hypothesis   The hypothesis that leads to knowledge of the distribution of the test statistic.

observed numbers   The values we obtain from conducting the experiment.

one-tailed tests   Tests of hypotheses where we will reject the hypothesis only if the values fall in one specified tail (either the right- or left-hand tail) of the normal (or t-) distribution when converted to z-scores (or t-scores).

options   The rights to buy or sell stock at some future time. They may be bought and sold, as the stock.

paired data   Two equal-sized sets of data that have a natural coupling.

parameter   A number derived from knowledge of the entire population, such as the *minimum* of the population, or the *mean* of the population.

percentile   One of the division points between 100 equal-sized pieces of the population when the population is arranged in numerical order. The 73rd percentile is the number such that 73% of the population is smaller and 27% of the population is larger.

pie chart   A circular figure divided into wedge-shaped sectors whose areas represent the relative proportions of the various subdivisions of the whole. See page 5.

population   Any complete set of objects, usually numbers.

population parameter   See *parameter*.

population proportion   The proportion of 1s, or "successes," in a binomial population.

positive correlation   A relationship in which two variables increase or decrease together. The marginal return is positive.

probability   Likelihood. The proportion or percentage of the time that an event occurs.

quality control   The statistical procedure used to keep a production process operating at a desired level or consistency.

quartile   One of the division points between four equal-sized pieces of the population when the population is arranged in numerical order. The second quartile is the number such that 2/4 of the population is smaller and 2/4 is larger.

quintile   One of the division points between five equal-sized pieces of the population when the population is arranged in numerical order. The third quintile is the number such that 3/5 of the population is smaller and 2/5 of the population is larger.

randomizing   Placing in random order.

random numbers   Numbers that are "chance" outcomes of some process. Usually, these are uniformly distributed random digits, meaning that each digit (0, 1, 2, 3, 4, 5, 6, 7, 8, or 9) records the result of a process of rolling a 10-sided *fair* die. The likelihood of any single digit is the same as that of any other digit; similarly for pairs, triples, and so on. See Table 8 at the back of the book.

random order   The result of ordering data in a random sequence, like shuffling cards.

random sample   A sample (subset) of a population that is selected on some "chance" basis, usually in such a way that every element of the population is as likely to be selected as any other element.

random variable   Function.

range   The largest number in a set minus the smallest.

rank-sum   A statistic based on ranking data. See *Mann–Whitney*.

rare event   Something that happens only very infrequently, for example, 5 times out of 100, or 1 time out of 100, or 1 time out of 1000.

regression coefficients   The constants in the regression equation.

regression equation   The equation of the regression line.

regression line   The straight line that "best fits" the given data; see *least squares*.

relative frequency   The number of occurrences of a given class, divided by the total number of numbers in the population; a percentage, or proportion, of the whole.

robust statistics   Statistical inference based on assumptions of normality, which holds true even when the assumptions fail to hold.

sample   A collection of numbers usually drawn at random from a population of numbers.

sample mean   The mean, or average, of a sample.

sample proportion   The proportion of 1s, or "successes," in a sample.

sample statistic   See *statistic*.

sampling without replacement   The process of sampling that does not allow the possibility that the same element is drawn twice from the population. See *sampling with replacement*.

sampling with replacement   The process of sampling that allows the possibility that the same element is drawn twice from the population. See *sampling without replacement*.

sigma   The standard deviation.

significance level   The proportion of the time that a hypothesis test will lead to a type 1 error, usually 0.10, 0.05, or 0.01.

sign test A nonparametric test based on replacing all observed values by either + or −, since the values observed are larger or smaller than some specified value, and then testing the resulting *binomial distribution*.

slope The coefficient of x in the regression equation.

small sample Usually less than 30.

standard deviation A measure of spread, applied either to a population or to a sample. See page 53. Also, the square root of variance.

standard normal distribution The normal distribution with mean 0 and variance 1.

statistic A number derived from observing a sample from a population of numbers.

tally Counting one for each number that falls in a given interval or class.

tally sheet A sheet of paper on which we record (tally) into which class each number in the population falls. The tally sheet records how many numbers in the population fall into each class.

t-distribution A distribution based on forming confidence intervals when we must estimate the population variance from the variance of a small sample. See Chapter 5.

three-sigma event An event that translates to a z-score of 3 or larger, or −3 or smaller. This can happen by chance roughly 1 time in 1000.

treatment Referring to the different populations in analysis of variance. Historically, the treatments were different fertilizers for the same crop.

treatment sum of squares In analysis of variance, the variation we observe from sample to sample, attributed to the effects of treatment.

trivariate data Data relating three variables. There are two independent and one dependent variable.

t-table A table of critical values of the *t*-distribution.

t-test A hypothesis test based on a small sample, leading to a *t*-score and examination of a critical value of the *t*-distribution. See Chapter 5.

two-sigma event An event that translates to a z-score of 2 or larger, or −2 or smaller. This can happen by chance roughly 5% of the time.

two-tailed test A test for which values can fall in either tail of the distribution and be cause for rejection of the null hypothesis.

two-way ANOVA The case when we form a model that has two different "treatments."

type 1 error Rejecting the null hypothesis when it is correct.

type 2 error Accepting the null hypothesis when it is false.

uncertainty principle The principle whereby for a fixed amount of effort, we can't both locate the parameter very precisely, and do so with a high degree of confidence. (The term is borrowed from physics, where it has another meaning.)

U-statistic   See *rank-sum*.

variability   See *variation*.

variance   The specific measure of variation that computes mean-squared deviations from the mean. The square of the standard deviation.

variance ratio   Same as *F*-ratio.

variation   The degree to which elements in a population fluctuate when sampled. See *variance*.

volatility   In referring to stock prices, the tendency to fluctuate over time.

z-**score**   The distance of the given value from the mean of the distribution or population, when measured in standard deviations.

Quick Symbol Guide

$\bar{x}$	Sample mean (pp. 18, 21, 98–103)
Σ	Summation sign; it means add up the objects that follow (p. 18)
$\sum_{i=1}^{15}$	Add up the 15 objects, numbered 1 to 15, that follow (pp. 18, 21)
$\boxed{\Sigma +}$	A special statistical key on many calculators (p. 18)
$\boxed{\text{2nd}}$	Choose the second function on any two-function key next struck (p. 19)
χ^2, 3 d.f.	Chi-square distribution, or chi-square statistic, with 3 degrees of freedom (pp. 153–155)
$F_{0.01}$ (2, 9)	The 1% critical value of the F-distribution, with 2 numerator degrees of freedom and 9 denominator degrees of freedom (pp. 199, 299)
σ	Standard deviation of the population (p. 53)
σ_{DSM} or $\sigma_{\bar{x}}$	Standard deviation of the distribution of sample means (p. 96)
s	Standard deviation of the sample (p. 53)
$\boxed{x^2}$	The function "squaring" (p. 48)
$\lvert A - B \rvert$	The distance between the numbers A and B (p. 47)
s^2	Sample variance (p. 51)
σ^2 or σ^2_{pop}	Population variance (p. 46)
5!	5 factorial $= 1 \times 2 \times 3 \times 4 \times 5 = 120$ (p. 75)
$\binom{5}{3}$	Binomial coefficient (p. 75)
$\cong$	"Is approximately equal to"

Quick Formula Guide

1. Sample mean: $\bar{x} = \dfrac{\Sigma x_i}{n} = \dfrac{x_1 + x_2 + x_3 + \cdots + x_n}{n}$ (pp. 18, 21)

2. Population variance: $\sigma^2 = \dfrac{\Sigma(x_i - \mu)^2}{n}$ (p. 46)

     Shortcut formula: $\sigma^2 = \dfrac{\Sigma(x_i^2) - \dfrac{(\Sigma x_i)^2}{n}}{n}$ (p. 58)

3. Sample variance: $s^2 = \dfrac{\Sigma(x_i - \bar{x})^2}{n - 1}$ (p. 51)

     Shortcut formula: $s^2 = \dfrac{\Sigma(x_i^2) - \dfrac{(\Sigma x_i)^2}{n}}{n - 1}$ (p. 58)

4. Standard deviation: $\sigma = \sqrt{\sigma^2}$ or $s = \sqrt{s^2}$

5. 95% confidence interval for the population mean; sample size $n \geq 30$:

     $\bar{x} \pm \dfrac{1.96s}{\sqrt{n}}$ (pp. 101–103)

6. z-scores: $z = \dfrac{x - \mu}{\sigma}$ or $z = \dfrac{x - \bar{x}}{s/\sqrt{n}}$ (pp. 66, 104)

7. Binomial distribution: probability of exactly k successes out of n trials, probability of success $= p$:

$$\binom{n}{k} p^k (1 - p)^{n-k} \quad \text{(p. 74)}$$

8. Standard deviation, differences between means:

$$\sigma_{\text{DDSM}} \cong \sqrt{\frac{s_A^2}{n_A} + \frac{s_B^2}{n_B}} \quad \text{(pp. 115–118)}$$

9. z-score, differences between means (large sample):

$$z = \frac{\bar{x}_A - \bar{x}_B}{\sqrt{\frac{s_A^2}{n_A} + \frac{s_B^2}{n_B}}}$$

is approximately standard normal   (p. 116)

10. t-score, differences between means (small sample):

$$t = \frac{x_A - \bar{x}_B}{\sqrt{\frac{\Sigma(x_A - \bar{x}_A)^2 + \Sigma(x_B - \bar{x}_B)^2}{n_A + n_B - 2} \left(\frac{1}{n_A} + \frac{1}{n_B} \right)}} \quad \text{(p. 182)}$$

11. Standard deviation, differences between proportions:

$$\sigma_{\text{DDSP}} \cong \sqrt{\frac{\bar{p}_A(1 - \bar{p}_A)}{n_A} + \frac{\bar{p}_B(1 - \bar{p}_B)}{n_B}} \quad \text{(p. 120)}$$

12. Contingency tables, n rows, m columns:

$$\chi^2 = \sum_{i=1}^{m \cdot n} \frac{(o_i - e_i)^2}{e_i}, \quad \text{with } (m - 1)(n - 1) \text{ d.f.} \quad \text{(p. 154)}$$

13. ANOVA table:

Source of Variation	d.f.	Sum of Squares	Mean Square	F-ratio
Treatment	$k - 1$	SS(Tr)	$\dfrac{\text{SS(Tr)}}{k - 1}$	$F = \dfrac{(\quad)}{(\quad)}$
Error	$N - k$	SSE	$\dfrac{\text{SSE}}{N - k}$	

where k = Number of populations;
   N = Total number of data;

$$\text{SS(Tr)} = \sum_{i=1}^{k} \frac{T_i^2}{n_i} - \frac{T^2}{N}; \qquad \text{SST} = \sum_{i,j} x_{ij}^2 - \frac{T^2}{N};$$

$$T_i = \sum_{j=1}^{n_i} x_{ij}; \quad T = \sum_{i=1}^{k} T_i; \qquad \text{SSE} = \text{SST} - \text{SS(Tr)} \quad \text{(p. 208)}$$

14. Mann–Whitney U-statistic (rank-sum):

R_A = Sum of the ranks of data from population A;

R_B = Sum of the ranks of data from population B:

$$R_B = \frac{n(n+1)}{2} - R_A \quad \text{or} \quad R_A = \frac{n(n+1)}{2} - R_B,$$

where $n = n_A + n_B$ = Sum of the sizes of sample A and sample B;

$$U_A = n_A n_B + \frac{n_A(n_A+1)}{2} - R_A$$

$$U_B = n_A n_B + \frac{n_B(n_B+1)}{2} - R_B$$

U = Smaller of U_A and U_B (p. 258)

15. Regression line: $y = mx + b$,

where $m = \dfrac{n(\Sigma xy) - (\Sigma x)(\Sigma y)}{n(\Sigma x^2) - (\Sigma x)^2}$

$b = \dfrac{(\Sigma x^2)(\Sigma y) - (\Sigma x)(\Sigma xy)}{n(\Sigma x^2) - (\Sigma x)^2}$ (p. 231)

16. Correlation: $r = \dfrac{n(\Sigma xy) - (\Sigma x)(\Sigma y)}{\sqrt{n(\Sigma x^2) - (\Sigma x)^2}\,\sqrt{n(\Sigma y^2) - (\Sigma y)^2}}$ (p. 235)

17. Estimation of σ $(n < 30)$: 95% confidence interval for σ^2 is

$$\frac{(n-1)s^2}{\chi^2_{0.025}} \leq \sigma^2 \leq \frac{(n-1)s^2}{\chi^2_{0.975}} \quad \text{(p. 278)}$$

18. Estimation of σ $(n \geq 30)$: s is approximately normal,

$$\text{Mean} = s; \quad \text{Standard deviation} = \frac{s}{\sqrt{2n}}$$

95% confidence interval for σ is $s \pm 1.96 \left(\dfrac{s}{\sqrt{2n}} \right)$ (p. 278)

Tables

Table 1   Areas of the Normal Distribution

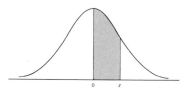

The entries in the table are the areas under the standard normal curve between 0 and z.

z	0.00	0.01	0.02	0.03	0.04	0.05	0.06	0.07	0.08	0.09
0.0	.0000	.0040	.0080	.0120	.0160	.0199	.0239	.0279	.0319	.0359
0.1	.0398	.0438	.0478	.0517	.0557	.0596	.0636	.0675	.0714	.0753
0.2	.0793	.0832	.0871	.0910	.0948	.0987	.1026	.1064	.1103	.1141
0.3	.1179	.1217	.1255	.1293	.1331	.1368	.1406	.1443	.1480	.1517
0.4	.1554	.1591	.1628	.1664	.1700	.1736	.1772	.1808	.1844	.1879
0.5	.1915	.1950	.1985	.2019	.2054	.2088	.2123	.2157	.2190	.2224
0.6	.2257	.2291	.2324	.2357	.2389	.2422	.2454	.2486	.2517	.2549
0.7	.2580	.2611	.2642	.2673	.2704	.2734	.2764	.2794	.2823	.2852
0.8	.2881	.2910	.2939	.2967	.2995	.3023	.3051	.3078	.3106	.3133
0.9	.3159	.3186	.3212	.3238	.3264	.3289	.3315	.3340	.3365	.3389
1.0	.3413	.3438	.3461	.3485	.3508	.3531	.3554	.3577	.3599	.3621
1.1	.3643	.3665	.3686	.3708	.3729	.3749	.3770	.3790	.3810	.3830
1.2	.3849	.3869	.3888	.3907	.3925	.3944	.3962	.3980	.3997	.4015
1.3	.4032	.4049	.4066	.4082	.4099	.4115	.4131	.4147	.4162	.4177
1.4	.4192	.4207	.4222	.4236	.4251	.4265	.4279	.4292	.4306	.4319
1.5	.4332	.4345	.4357	.4370	.4382	.4394	.4406	.4418	.4429	.4441
1.6	.4452	.4463	.4474	.4484	.4495	.4505	.4515	.4525	.4535	.4545
1.7	.4554	.4564	.4573	.4582	.4591	.4599	.4608	.4616	.4625	.4633
1.8	.4641	.4649	.4656	.4664	.4671	.4678	.4686	.4693	.4699	.4706
1.9	.4713	.4719	.4726	.4732	.4738	.4744	.4750	.4756	.4761	.4767
2.0	.4772	.4778	.4783	.4788	.4793	.4798	.4803	.4808	.4812	.4817
2.1	.4821	.4826	.4830	.4834	.4838	.4842	.4846	.4850	.4854	.4857
2.2	.4861	.4864	.4868	.4871	.4875	.4878	.4881	.4884	.4887	.4890
2.3	.4893	.4896	.4898	.4901	.4904	.4906	.4909	.4911	.4913	.4916
2.4	.4918	.4920	.4922	.4925	.4927	.4929	.4931	.4932	.4934	.4936
2.5	.4938	.4940	.4941	.4943	.4945	.4946	.4948	.4949	.4951	.4952
2.6	.4953	.4955	.4956	.4957	.4959	.4960	.4961	.4962	.4963	.4964
2.7	.4965	.4966	.4967	.4968	.4969	.4970	.4971	.4972	.4973	.4974
2.8	.4974	.4975	.4976	.4977	.4977	.4978	.4979	.4979	.4980	.4981
2.9	.4981	.4982	.4982	.4983	.4984	.4984	.4985	.4985	.4986	.4986
3.0	.4987	.4987	.4987	.4988	.4988	.4989	.4989	.4989	.4990	.4990
4.0	.49997									
5.0	.4999997									
6.0	.499999999									

TABLE 2   **295**

Table 2   Normal Table for Confidence Intervals

When you go out ± this many sigmas from the mean of the normal distribution (k)	You capture this percentage of the normal distribution (P)
0.00	0.0%
0.10	8.0
0.20	15.9
0.30	23.6
0.40	31.1
0.50	38.3
0.60	45.1
0.70	51.6
0.80	57.6
0.90	63.2
1.00	**68.3**
1.10	72.9
1.20	77.0
1.30	80.6
1.40	83.8
1.50	86.6
1.60	89.0
1.645	**90.0**
1.70	91.1
1.80	92.8
1.90	94.3
1.96	**95.0**
2.00	95.4
2.10	96.4
2.20	97.2
2.30	97.9
2.40	98.4
2.50	98.8
2.576	**99.0**
2.60	99.1
2.70	99.3
2.80	99.5
2.90	99.6
3.00	**99.7**
4.00	99.994
5.00	99.99994
6.00	99.9999998

Table 3   Critical Values of the *t*-Distribution

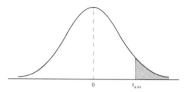

Each entry in the table under $t_{0.050}$ with given number of degrees of freedom (d.f.) is the value along the number line such that 0.05, or 5%, of the total area under the curve lies to the right. This is the 5% critical value of *t*. The table gives 10%, 5%, 2½%, 1%, and ½% critical values.

n	d.f.	$t_{0.100}$	$t_{0.050}$	$t_{0.025}$	$t_{0.010}$	$t_{0.005}$	d.f.
2	1	3.078	6.314	12.706	31.821	63.657	1
3	2	1.886	2.920	4.303	6.965	9.925	2
4	3	1.638	2.353	3.182	4.541	5.841	3
5	4	1.533	2.132	2.776	3.747	4.604	4
6	5	1.476	2.015	2.571	3.365	4.032	5
7	6	1.440	1.943	2.447	3.143	3.707	6
8	7	1.415	1.895	2.365	2.998	3.499	7
9	8	1.397	1.860	2.306	2.896	3.355	8
10	9	1.383	1.833	2.262	2.821	3.250	9
11	10	1.372	1.812	2.228	2.764	3.169	10
12	11	1.363	1.796	2.201	2.718	3.106	11
13	12	1.356	1.782	2.179	2.681	3.055	12
14	13	1.350	1.771	2.160	2.650	3.012	13
15	14	1.345	1.761	2.145	2.624	2.977	14
16	15	1.341	1.753	2.131	2.602	2.947	15
17	16	1.337	1.746	2.120	2.583	2.921	16
18	17	1.333	1.740	2.110	2.567	2.898	17
19	18	1.330	1.734	2.101	2.552	2.878	18
20	19	1.328	1.729	2.093	2.539	2.861	19
21	20	1.325	1.725	2.086	2.528	2.845	20
22	21	1.323	1.721	2.080	2.518	2.831	21
23	22	1.321	1.717	2.074	2.508	2.819	22
24	23	1.319	1.714	2.069	2.500	2.807	23
25	24	1.318	1.711	2.064	2.492	2.797	24
26	25	1.316	1.708	2.060	2.485	2.787	25
27	26	1.315	1.706	2.056	2.479	2.779	26
28	27	1.314	1.703	2.052	2.473	2.771	27
29	28	1.313	1.701	2.048	2.467	2.763	28
30	29	1.311	1.699	2.045	2.462	2.756	29
∞	∞	1.282	1.645	1.960	2.326	2.576	∞

TABLE 4   **297**

Table 4   Critical Values of the Chi-Square Distribution

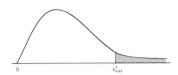

Each entry in the table under $\chi^2_{0.05}$ with given number of degrees of freedom (d.f.) is the value along the number line such that 0.05, or 5%, of the total area under the curve lies to the right. This is the 5% critical value of chi-square (χ^2). The table gives 0.995, 0.99, 0.975, 0.95, 0.05, 0.025, 0.01, and 0.005 critical values.

d.f.	$\chi^2_{0.995}$	$\chi^2_{0.99}$	$\chi^2_{0.975}$	$\chi^2_{0.95}$	$\chi^2_{0.05}$	$\chi^2_{0.025}$	$\chi^2_{0.01}$	$\chi^2_{0.005}$	d.f.
1	.0000393	.000157	.000982	.00393	3.841	5.024	6.635	7.879	1
2	.0100	.0201	.0506	.103	5.991	7.378	9.210	10.597	2
3	.0717	.115	.216	.352	7.815	9.348	11.345	12.838	3
4	.207	.297	.484	.711	9.488	11.143	13.277	14.860	4
5	.412	.554	.831	1.145	11.070	12.832	15.086	16.750	5
6	.676	.872	1.237	1.635	12.592	14.449	16.812	18.548	6
7	.989	1.239	1.690	2.167	14.067	16.013	18.475	20.278	7
8	1.344	1.646	2.180	2.733	15.507	17.535	20.090	21.955	8
9	1.735	2.088	2.700	3.325	16.919	19.023	21.666	23.589	9
10	2.156	2.558	3.247	3.940	18.307	20.483	23.209	25.188	10
11	2.603	3.053	3.816	4.575	19.675	21.920	24.725	25.757	11
12	3.074	3.571	4.404	5.226	21.026	23.337	26.217	28.300	12
13	3.565	4.107	5.009	5.892	22.362	24.736	27.688	29.819	13
14	4.075	4.660	5.629	6.571	23.685	26.119	29.141	31.319	14
15	4.601	5.229	6.262	7.261	24.996	27.488	30.578	32.801	15
16	5.142	5.812	6.908	7.962	26.296	28.845	32.000	34.267	16
17	5.697	6.408	7.564	8.672	27.587	30.191	33.409	35.718	17
18	6.265	7.015	8.231	9.390	28.869	31.526	34.805	37.156	18
19	6.844	7.633	8.907	10.117	30.144	32.852	36.191	38.582	19
20	7.434	8.260	9.591	10.851	31.410	34.170	37.566	39.997	20
21	8.034	8.897	10.283	11.591	32.671	35.479	38.932	41.401	21
22	8.643	9.542	10.982	12.338	33.924	36.781	40.289	42.796	22
23	9.260	10.196	11.689	13.091	35.172	38.076	41.638	44.181	23
24	9.886	10.856	12.401	13.848	36.415	39.364	42.980	45.558	24
25	10.520	11.524	13.120	14.611	37.652	40.646	44.314	46.928	25
26	11.160	12.198	13.844	15.379	38.885	41.923	45.642	48.290	26
27	11.808	12.879	14.573	16.151	40.113	43.194	46.963	49.645	27
28	12.461	13.565	15.308	16.928	41.337	44.461	48.278	50.993	28
29	13.121	14.256	16.047	17.708	42.557	45.722	49.588	52.336	29
30	13.787	14.953	16.791	18.493	43.773	46.979	50.892	53.672	30

Based on Table 8 of *Biometrika Tables for Statisticians, Vol. I* (Cambridge University Press, 1954), by permission of the *Biometrika* trustees.

Table 5a 0.05 Critical Values of F

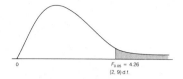

$F_{0.05} = 4.26$
(2, 9) d.f.

Each entry in the table is a 5%, or 0.05, critical value of the F-distribution for the given numerator d.f. and denominator d.f. The entry gives the value along the number line such that 5% of the area of the distribution lies to the right. [As the figure indicates, with (2, 9) d.f., $F_{0.05} = 4.26$.]

Degrees of Freedom for Numerator

	1	2	3	4	5	6	7	8	9	10	12	15	20	24	30	40	60	120	∞
1	161	200	216	225	230	234	237	239	241	242	244	246	248	249	250	251	252	253	254
2	18.5	19.0	19.2	19.2	19.3	19.3	19.4	19.4	19.4	19.4	19.4	19.4	19.4	19.4	19.5	19.5	19.5	19.5	19.5
3	10.1	9.55	9.28	9.12	9.01	8.94	8.89	8.85	8.81	8.79	8.74	8.70	8.66	8.64	8.62	8.59	8.57	8.55	8.53
4	7.71	6.94	6.59	6.39	6.26	6.16	6.09	6.04	6.00	5.96	5.91	5.86	5.80	5.77	5.75	5.72	5.69	5.66	5.63
5	6.61	5.79	5.41	5.19	5.05	4.95	4.88	4.82	4.77	4.74	4.68	4.62	4.56	4.53	4.50	4.46	4.43	4.40	4.37
6	5.99	5.14	4.76	4.53	4.39	4.28	4.21	4.15	4.10	4.06	4.00	3.94	3.87	3.84	3.81	3.77	3.74	3.70	3.67
7	5.59	4.74	4.35	4.12	3.97	3.87	3.79	3.73	3.68	3.64	3.57	3.51	3.44	3.41	3.38	3.34	3.30	3.27	3.23
8	5.32	4.46	4.07	3.84	3.69	3.58	3.50	3.44	3.39	3.35	3.28	3.22	3.15	3.12	3.08	3.04	3.01	2.97	2.93
9	5.12	4.26	3.86	3.63	3.48	3.37	3.29	3.23	3.18	3.14	3.07	3.01	2.94	2.90	2.86	2.83	2.79	2.75	2.71
10	4.96	4.10	3.71	3.48	3.33	3.22	3.14	3.07	3.02	2.98	2.91	2.85	2.77	2.74	2.70	2.66	2.62	2.58	2.54
11	4.84	3.98	3.59	3.36	3.20	3.09	3.01	2.95	2.90	2.85	2.79	2.72	2.65	2.61	2.57	2.53	2.49	2.45	2.40
12	4.75	3.89	3.49	3.26	3.11	3.00	2.91	2.85	2.80	2.75	2.69	2.62	2.54	2.51	2.47	2.43	2.38	2.34	2.30
13	4.67	3.81	3.41	3.18	3.03	2.92	2.83	2.77	2.71	2.67	2.60	2.53	2.46	2.42	2.38	2.34	2.30	2.25	2.21
14	4.60	3.74	3.34	3.11	2.96	2.85	2.76	2.70	2.65	2.60	2.53	2.46	2.39	2.35	2.31	2.27	2.22	2.18	2.13
15	4.54	3.68	3.29	3.06	2.90	2.79	2.71	2.64	2.59	2.54	2.48	2.40	2.33	2.29	2.25	2.20	2.16	2.11	2.07
16	4.49	3.63	3.24	3.01	2.85	2.74	2.66	2.59	2.54	2.49	2.42	2.35	2.28	2.24	2.19	2.15	2.11	2.06	2.01
17	4.45	3.59	3.20	2.96	2.81	2.70	2.61	2.55	2.49	2.45	2.38	2.31	2.23	2.19	2.15	2.10	2.06	2.01	1.96
18	4.41	3.55	3.16	2.93	2.77	2.66	2.58	2.51	2.46	2.41	2.34	2.27	2.19	2.15	2.11	2.06	2.02	1.97	1.92
19	4.38	3.52	3.13	2.90	2.74	2.63	2.54	2.48	2.42	2.38	2.31	2.23	2.16	2.11	2.07	2.03	1.98	1.93	1.88
20	4.35	3.49	3.10	2.87	2.71	2.60	2.51	2.45	2.39	2.35	2.28	2.20	2.12	2.08	2.04	1.99	1.95	1.90	1.84
21	4.32	3.47	3.07	2.84	2.68	2.57	2.49	2.42	2.37	2.32	2.25	2.18	2.10	2.05	2.01	1.96	192	1.87	1.81
22	4.30	3.44	3.05	2.82	2.66	2.55	2.46	2.40	2.34	2.30	2.23	2.15	2.07	2.03	1.98	1.94	1.89	1.84	1.78
23	4.28	3.42	3.03	2.80	2.64	2.53	2.44	2.37	2.32	2.27	2.20	2.13	2.05	2.01	1.96	1.91	1.86	1.81	1.76
24	4.26	3.40	3.01	2.78	2.62	2.51	2.42	2.36	2.30	2.25	2.18	2.11	2.03	1.98	1.94	1.89	1.84	1.79	1.73
25	4.24	3.39	2.99	2.76	2.60	2.49	2.40	2.34	2.28	2.24	2.16	2.09	2.01	1.96	1.92	1.87	1.82	1.77	1.71
30	4.17	3.32	2.92	2.69	2.53	2.42	2.33	2.27	2.21	2.16	2.09	2.01	1.93	1.89	1.84	1.79	1.74	1.68	1.62
40	4.08	3.23	2.84	2.61	2.45	2.34	2.25	2.18	2.12	2.08	2.00	1.92	1.84	1.79	1.74	1.69	1.64	1.58	1.51
60	4.00	3.15	2.76	2.53	2.37	2.25	2.17	2.10	2.04	1.99	1.92	1.84	1.75	1.70	1.65	1.59	1.53	1.47	1.39
120	3.92	3.07	2.68	2.45	2.29	2.18	2.09	2.02	1.96	1.91	1.83	1.75	1.66	1.61	1.55	1.50	1.43	1.35	1.25
∞	3.84	3.00	2.60	2.37	2.21	2.10	2.01	1.94	1.88	1.83	1.75	1.67	1.57	1.52	1.46	1.39	1.32	1.22	1.00

Degrees of Freedom for Denominator

Table 5b   0.01 Critical Values of F

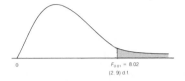

$F_{.01} = 8.02$
(2, 9) d.f.

Each entry in the table is a 1%, or 0.01, critical value of the F-distribution for the given numerator d.f. and denominator d.f. The entry gives the value along the number line such that 1% of the area of the distribution lies to the right. [As the figure indicates, with (2, 9) d.f., $F_{0.01} = 8.02$.]

Degrees of Freedom for Numerator

		1	2	3	4	5	6	7	8	9	10	12	15	20	24	30	40	60	120	∞
Degrees of Freedom for Denominator	1	4,052	5,000	5,403	5,625	5,764	5,859	5,928	5,982	6,023	6,056	6,106	6,157	6,209	6,235	6,261	6,287	6,313	6,339	6,366
	2	98.5	99.0	99.2	99.2	99.3	99.3	99.4	99.4	99.4	99.4	99.4	99.4	99.4	99.5	99.5	99.5	99.5	99.5	99.5
	3	34.1	30.8	29.5	28.7	28.2	27.9	27.7	27.5	27.3	27.2	27.1	26.9	26.7	26.6	26.5	26.4	26.3	26.2	26.1
	4	21.2	18.0	16.7	16.0	15.5	15.2	15.0	14.8	14.7	14.5	14.4	14.2	14.0	13.9	13.8	13.7	13.7	13.6	13.5
	5	16.3	13.3	12.1	11.4	11.0	10.7	10.5	10.3	10.2	10.1	9.89	9.72	9.55	9.47	9.38	9.29	9.20	9.11	9.02
	6	13.7	10.9	9.78	9.15	8.75	8.47	8.26	8.10	7.98	7.87	7.72	7.56	7.40	7.31	7.23	7.14	7.06	6.97	6.88
	7	12.2	9.55	8.45	7.85	7.46	7.19	6.99	6.84	6.72	6.62	6.47	6.31	6.16	6.07	5.99	5.91	5.82	5.74	5.65
	8	11.3	8.65	7.59	7.01	6.63	6.37	6.18	6.03	5.91	5.81	5.67	5.52	5.36	5.28	5.20	5.12	5.03	4.95	4.86
	9	10.6	8.02	6.99	6.42	6.06	5.80	5.61	5.47	5.35	5.26	5.11	4.96	4.81	4.73	4.65	4.57	4.48	4.40	4.31
	10	10.0	7.56	6.55	5.99	5.64	5.39	5.20	5.06	4.94	4.85	4.71	4.56	4.41	4.33	4.25	4.17	4.08	4.00	3.91
	11	9.65	7.21	6.22	5.67	5.32	5.07	4.89	4.74	4.63	4.54	4.40	4.25	4.10	4.02	3.94	3.86	3.78	3.69	3.60
	12	9.33	6.93	5.95	5.41	5.06	4.82	4.64	4.50	4.39	4.30	4.16	4.01	3.86	3.78	3.70	3.62	3.54	3.45	3.36
	13	9.07	6.70	5.74	5.21	4.86	4.62	4.44	4.30	4.19	4.10	3.96	3.82	3.66	3.59	3.51	3.43	3.34	3.25	3.17
	14	8.86	6.51	5.56	5.04	4.70	4.46	4.28	4.14	4.03	3.94	3.80	3.66	3.51	3.43	3.35	3.27	3.18	3.09	3.00
	15	8.68	6.36	5.42	4.89	4.56	4.32	4.14	4.00	3.89	3.80	3.67	3.52	3.37	3.29	3.21	3.13	3.05	2.96	2.87
	16	8.53	6.23	5.29	4.77	4.44	4.20	4.03	3.89	3.78	3.69	3.55	3.41	3.26	3.18	3.10	3.02	2.93	2.84	2.75
	17	8.40	6.11	5.19	4.67	4.34	4.10	3.93	3.79	3.68	3.59	3.46	3.31	3.16	3.08	3.00	2.92	2.83	2.75	2.65
	18	8.29	6.01	5.09	4.58	4.25	4.01	3.84	3.71	3.60	3.51	3.37	3.23	3.08	3.00	2.92	2.84	2.75	2.66	2.57
	19	8.19	5.93	5.01	4.50	4.17	3.94	3.77	3.63	3.52	3.43	3.30	3.15	3.00	2.92	2.84	2.76	2.67	2.58	2.49
	20	8.10	5.85	4.94	4.43	4.10	3.87	3.70	3.56	3.46	3.37	3.23	3.09	2.94	2.86	2.78	2.69	2.61	2.52	2.42
	21	8.02	5.78	4.87	4.37	4.04	3.81	3.64	3.51	3.40	3.31	3.17	3.03	2.88	2.80	2.72	2.64	2.55	2.46	2.36
	22	7.95	5.72	4.82	4.31	3.99	3.76	3.59	3.45	3.35	3.26	3.12	2.98	2.83	2.75	2.67	2.58	2.50	2.40	2.31
	23	7.88	5.66	4.76	4.26	3.94	3.71	3.54	3.41	3.30	3.21	3.07	2.93	2.78	2.70	2.62	2.54	2.45	2.35	2.26
	24	7.82	5.61	4.72	4.22	3.90	3.67	3.50	3.36	3.26	3.17	3.03	2.89	2.74	2.66	2.58	2.49	2.40	2.31	2.21
	25	7.77	5.57	4.68	4.18	3.86	3.63	3.46	3.32	3.22	3.13	2.99	2.85	2.70	2.62	2.53	2.45	2.36	2.27	2.17
	30	7.56	5.39	4.51	4.02	3.70	3.47	3.30	3.17	3.07	2.98	2.84	2.70	2.55	2.47	2.39	2.30	2.21	2.11	2.01
	40	7.31	5.18	4.31	3.83	3.51	3.29	3.12	2.99	2.89	2.80	2.66	2.52	2.37	2.29	2.20	2.11	2.02	1.92	1.80
	60	7.08	4.98	4.13	3.65	3.34	3.12	2.95	2.82	2.72	2.63	2.50	2.35	2.20	2.12	2.03	1.94	1.84	1.73	1.60
	120	6.85	4.79	3.95	3.48	3.17	2.96	2.79	2.66	2.56	2.47	2.34	2.19	2.03	1.95	1.86	1.76	1.66	1.53	1.38
	∞	6.63	4.61	3.78	3.32	3.02	2.80	2.64	2.51	2.41	2.32	2.18	2.04	1.88	1.79	1.70	1.59	1.47	1.32	1.00

Table 6a   95% Confidence Intervals for Proportions

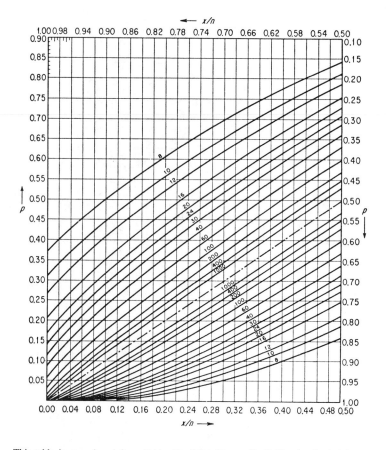

This table is reproduced from Table 41 of the *Biometrika Tables for Statisticians*, Vol. I (New York: Cambridge University Press, 1954), by permission of the *Biometrika* trustees.

Table 6b   99% Confidence Intervals for Proportions

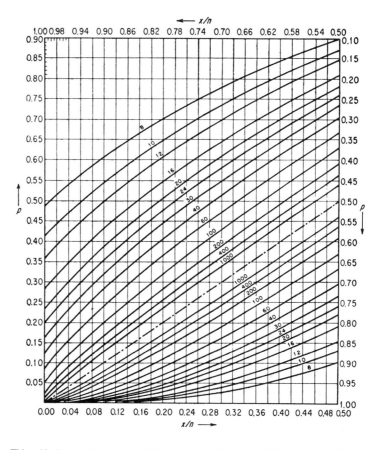

This table is reproduced from Table 41 of the *Biometrika Tables for Statisticians*, Vol. I (New York: Cambridge University Press, 1954), by permission of the *Biometrika* trustees.

Table 7a   Critical Values of the Binomial Distribution, $p = \frac{1}{2}$
n from 5 to 15

We flip a fair coin n times and we observe k "heads."

n	One-Tailed Tests	Two-Tailed Tests
At the 0.01 level of significance, reject $p = \frac{1}{2}$ if:		
$n = 5$	Cannot reject	Cannot reject
$n = 6$	Cannot reject	Cannot reject
$n = 7$	$k = 0$ or $k = 7$	Cannot reject
$n = 8$	$k = 0$ or $k = 8$	$k = 0$ or $k = 8$
$n = 9$	$k = 0$ or $k = 9$	$k = 0$ or $k = 9$
$n = 10$	$k = 0$ or $k = 10$	$k = 0$ or $k = 10$
$n = 11$	$k \leq 1$ or $k \geq 10$	$k \leq 1$ or $k \geq 10$
$n = 12$	$k \leq 1$ or $k \geq 11$	$k \leq 1$ or $k \geq 11$
$n = 13$	$k \leq 1$ or $k \geq 12$	$k \leq 1$ or $k \geq 12$
$n = 14$	$k \leq 2$ or $k \geq 12$	$k \leq 1$ or $k \geq 13$
$n = 15$	$k \leq 2$ or $k \geq 13$	$k \leq 2$ or $k \geq 13$
At the 0.05 level of significance, reject $p = \frac{1}{2}$ if:		
$n = 5$	$k = 0$ or $k = 5$	Cannot reject
$n = 6$	$k = 0$ or $k = 6$	$k = 0$ or $k = 6$
$n = 7$	$k = 0$ or $k = 7$	$k = 0$ or $k = 7$
$n = 8$	$k \leq 1$ or $k \geq 7$	$k = 0$ or $k = 8$
$n = 9$	$k \leq 1$ or $k \geq 8$	$k \leq 1$ or $k \geq 8$
$n = 10$	$k \leq 1$ or $k \geq 9$	$k \leq 1$ or $k \geq 9$
$n = 11$	$k \leq 2$ or $k \geq 9$	$k \leq 1$ or $k \geq 10$
$n = 12$	$k \leq 2$ or $k \geq 10$	$k \leq 2$ or $k \geq 10$
$n = 13$	$k \leq 3$ or $k \geq 10$	$k \leq 2$ or $k \geq 11$
$n = 14$	$k \leq 3$ or $k \geq 11$	$k \leq 2$ or $k \geq 12$
$n = 15$	$k \leq 3$ or $k \geq 12$	$k \leq 3$ or $k \geq 12$

Table 7b   Critical Values of the Binomial Distribution, $p = \frac{1}{2}$
n greater than 15

Flip n times. Observe k heads or k tails.

For n greater than 15, we use the normal approximation to the binomial distribution, with continuity correction, as follows: After observing n flips, take k to be the larger of the number of heads observed, or the number of tails observed. Thus, $2k \geq n$. Compute the test statistic

$$z = \frac{2k - 1 - n}{\sqrt{n}}$$

which, if $p = \frac{1}{2}$, is normally distributed with mean 0 and variance 1. Then if z is greater than or equal to the appropriate entry in the table below, reject the hypothesis $p = \frac{1}{2}$.

	For one-tailed tests	For two-tailed tests
At the 0.01 level of significance	2.33	2.576
At the 0.05 level of significance	1.645	1.96

304 TABLE 8

Table 8   A Table of Random Numbers

17748	04963	54859	88698	41755	56216	66852
41699	11732	17173	51865	09836	73966	65711
79589	95295	72895	40300	08852	27528	84648
32988	10194	94917	02760	28625	70476	76410
33047	03577	62599	78450	26245	91763	73117
98693	18728	94741	50252	56911	62693	73817
44806	15592	71357	07929	66728	47761	81472
51257	89555	75520	09030	39605	87507	85446
21795	38894	58070	56670	88445	85799	76200
64741	64336	95103	48140	13583	94911	13318
94242	32063	45233	36764	86132	12463	28385
65145	28152	39087	14351	71381	28133	68269
42446	08882	27067	81276	00835	63835	87174
24654	77371	26409	55524	86088	00069	59254
23937	90740	16866	78852	65889	32719	13758
32050	52052	24004	11861	69032	51915	23510
06732	27510	33761	67699	01009	07050	73324
63124	48061	59412	50064	39500	17450	18030
08317	27324	72723	93126	17700	94400	76075
15650	29970	95877	01657	92602	41043	05686
28491	03845	11507	13800	76690	75133	60456
69876	86563	61729	98135	42870	48578	29036
77457	79969	11339	08313	99293	00990	13595
54330	22406	86253	90974	83965	62732	85161
17301	70975	99129	33273	61993	88407	69399
19246	88097	44926	94620	27963	96478	21559
56895	04232	59604	60947	60775	73181	43264
20865	91683	80688	27499	53523	63110	57106
95353	44662	59433	01603	23156	89223	43429
70385	45863	75971	00815	01552	06392	31437
68723	47830	63010	83844	90942	74857	52419
57215	08409	81906	06626	10042	93629	37609
29793	37457	59377	56760	63348	24949	11859
14230	62887	92683	64416	29934	00755	09418
19096	96970	75917	63569	17906	38076	32135
23791	60249	83010	22693	35089	72994	04252
91382	45114	20245	43413	59744	01275	71326
04851	18280	14039	09224	78530	50566	49965
63558	09665	22610	67625	34683	03142	74733
91579	26023	81076	86874	12549	98699	54952
13193	33905	66936	54548	49505	62515	63903
40699	10396	81827	73236	66167	49728	03581
59148	95154	72852	15220	66319	13543	14071
38313	34016	18671	16151	08029	36954	03891
55431	90793	62603	43635	84249	88984	80993
51362	79907	77364	30193	42776	85611	57635
43996	73122	88474	37430	45246	11400	20986
70794	01041	74867	88312	93047	12088	86937
13168	31553	67891	98995	58159	04700	90443
00880	82899	66065	51734	20849	70198	67906

TABLE 9   **305**

Table 9   Critical Values of the *U*-Distribution

n_1 and n_2 are the sample sizes. Arrange to choose which is which, so that $n_1 \leq n_2$. Then reject the hypothesis that the two samples came from identical populations if the value of the statistic U is *less than or equal to* the indicated value in the table; these are two-sided tests, as U is chosen the smaller of two numbers. This condenses the table. If no entry appears where you look, check that $n_1 \leq n_2$. If so, then you cannot reject. If $n_1 > n_2$, then reverse n_1 and n_2.

0.05 Level of Significance

n_1 \ n_2	4	5	6	7	8	9	10
2					0	0	0
3		0	1	1	2	2	3
4	0	1	2	3	4	4	5
5		2	3	5	6	7	8
6			5	6	8	10	11
7				8	10	12	14
8					13	15	17
9						17	20
10							23

0.01 Level of Significance

n_1 \ n_2	5	6	7	8	9	10
3					0	0
4		0	0	1	1	2
5	0	1	1	2	3	4
6		2	3	4	5	6
7			4	6	7	9
8				7	9	11
9					11	13
10						16

Multiple Regression Computer Program

```
10    PRINT "INPUT NUMBER OF INDEPENDENT
      VARIABLES.": PRINT "FOR 3-VARIABLE
       DATA THIS NUMBER IS 2, ETC.": PRINT
      "THERE IS ALWAYS 1 DEPENDENT VARIA
      BLE"
20    INPUT N: DIM X(N + 2)
30    DIM S(N + 2,N + 2),Y(N + 2),Z(N + 2
      ),W(N + 2,N + 2)
35    FOR I = 1 TO N + 2: FOR K = 1 TO N +
       2
36 W(I,K) = 0:S(I,K) = 0: NEXT K:Y(I) =
      0:Z(I) = 0: NEXT I
40  J = 1
50    PRINT "INPUT OBSERVATION ";J;":": PRINT
      ""
51    FOR I = 1 TO N: PRINT "X(";I;") = "

55    INPUT X(I): NEXT I:X(N + 1) = 1
80    FOR I = 1 TO N + 1: FOR K = 1 TO N +
       1
81 W(I,K) = W(I,K) + X(I) * X(K): NEXT
      K: NEXT I
90    REM TO HERE WE HAVE THE CROSS PRODU
      CTS
100   PRINT "INPUT OBSERVATION NUMBER ";
      J;" OF THE DEPENDENT VARIABLE."
110   INPUT Y
120   FOR I = 1 TO N + 1:Y(I) = Y(I) + Y
       * X(I): NEXT I
130   PRINT "ANY MORE DATA? (PRESS 'Y' O
      R 'N')"
140   INPUT C$: IF C$ = "N" THEN 170
150   IF C$ < > "Y" THEN 130
160 J = J + 1: GOTO 50
170   REM  TO HERE ALL SUMS ACCUMULATED
171 JJ = J
175   FOR I = 1 TO N + 2: FOR K = 1 TO N
      + 2:S(I,K) = W(I,K): NEXT K: NEXT
      I
180   FOR I = 1 TO N + 1:S(I,N + 2) = Y(
      I): NEXT I
500 J = 1
510   FOR I = J TO N + 1: IF S(I,J) < >
      0 THEN 525
515   NEXT I
```

```
520   PRINT "MATRIX IS SINGULAR.": PRINT
      "IT CANNOT BE SOLVED.": PRINT "ENT
      ER MORE DATA."
521 J = JJ + 1: GOTO 50
525 V = S(I,J): FOR K = J TO N + 2:S(I,
      K) = S(I,K) / V: NEXT K: REM NORMA
      LIZE I-TH ROW
530   IF I = J THEN  GOTO 550
535   FOR K = J TO N + 2:T = S(I,K):S(I,
      K) = S(J,K):S(J,K) = T: NEXT K
550   REM TO HERE THE J-TH --TOP-- ROW I
      S NORMALIZED -- S(J,J)=1
555   IF J = N + 1 THEN 610: REM  ALL DO
      NE
560   FOR L = J + 1 TO N + 1: REM  ZERO
      LOWER ROWS
570 X = S(L,J): FOR K = J TO N + 2
590 S(L,K) = S(L,K) - X * S(J,K): NEXT
      K
600   NEXT L
610   REM  TO HERE ALL BELOW S(J,J) IS 0

620   IF J < N + 1 THEN J = J + 1: GOTO
      510
630   REM  TO HERE IN UPPER TRIANGULAR F
      ORM
650   Z(N + 1) = S(N + 1,N + 2)
655   FOR I = N TO 1 STEP  - 1
660 R = 0: FOR J = N + 1 TO I + 1 STEP
       - 1
670 R = R + S(I,J) * Z(J): NEXT J
680 Z(I) = S(I,N + 2) - R: NEXT I
690   REM  Z HAS THE COEFFICIENTS
694   PRINT "": PRINT ""
695   PRINT "REGRESSION EQUATION:


                          Y =  A1*X1 + A
      2*X2 + ...+ AN*XN + A0         "
698   FOR I = 1 TO N: PRINT "A";I;"=";Z(
      I): NEXT I: PRINT "A0=";Z(N + 1)
699   PRINT "": PRINT "DO YOU WANT TO PR
      EDICT WITH YOUR REGRESSION EQUATIO
      N? (PRESS 'Y' OR 'N')"
700   INPUT C$: IF C$ = "Y" THEN 750
701   IF C$ = "N" THEN  STOP
702   PRINT "PRESS 'Y' OR 'N'": GOTO 700

750   PRINT "": PRINT "INPUT THE VALUES
      OF YOUR VARIABLES:": PRINT ""
755   FOR I = 1 TO N: PRINT "X(";I;")=":
      INPUT X(I): NEXT I
760 SS = Z(N + 1): FOR I = 1 TO N:SS =
      SS + Z(I) * X(I): NEXT I
770   PRINT "": PRINT "Y PREDICTED = ";S
      S
775   GOTO 699
```

Example Index

Index

Quick Formula Guide

1. Sample mean: $\bar{x} = \dfrac{\Sigma x_i}{n} = \dfrac{x_1 + x_2 + x_3 + \cdots + x_n}{n}$ (pp. 18, 21)

2. Population variance: $\sigma^2 = \dfrac{\Sigma(x_i - \mu)^2}{n}$ (p. 46)

   Shortcut formula: $\sigma^2 = \dfrac{\Sigma(x_i^2) - \dfrac{(\Sigma x_i)^2}{n}}{n}$ (p. 58)

3. Sample variance: $s^2 = \dfrac{\Sigma(x_i - \bar{x})^2}{n - 1}$ (p. 51)

   Shortcut formula: $s^2 = \dfrac{\Sigma(x_i^2) - \dfrac{(\Sigma x_i)^2}{n}}{n - 1}$ (p. 58)

4. Standard deviation: $\sigma = \sqrt{\sigma^2}$ or $s = \sqrt{s^2}$

5. 95% confidence interval for the population mean; sample size $n \geqslant 30$:
   $\bar{x} \pm \dfrac{1.96s}{\sqrt{n}}$ (pp. 101–103)

6. z-scores: $z = \dfrac{x - \mu}{\sigma}$ or $z = \dfrac{x - \bar{x}}{s/\sqrt{n}}$ (pp. 66, 104)

7. Binomial distribution: probability of exactly k successes out of n trials, probability of success $= p$:

$$\binom{n}{k} p^k (1 - p)^{n-k} \quad \text{(p. 74)}$$

8. Standard deviation, differences between means:

$$\sigma_{\text{DDSM}} \cong \sqrt{\frac{s_A^2}{n_A} + \frac{s_B^2}{n_B}} \quad \text{(pp. 115–118)}$$

9. z-score, differences between means (large sample):

$$z = \frac{\bar{x}_A - \bar{x}_B}{\sqrt{\dfrac{s_A^2}{n_A} + \dfrac{s_B^2}{n_B}}}$$

is approximately standard normal (p. 116)

10. t-score, differences between means (small sample):

$$t = \frac{x_A - \bar{x}_B}{\sqrt{\dfrac{\Sigma(x_A - \bar{x}_A)^2 + \Sigma(x_B - \bar{x}_B)^2}{n_A + n_B - 2} \left(\dfrac{1}{n_A} + \dfrac{1}{n_B} \right)}} \quad \text{(p. 182)}$$

11. Standard deviation, differences between proportions:

$$\sigma_{\text{DDSP}} \cong \sqrt{\frac{\bar{p}_A(1 - \bar{p}_A)}{n_A} + \frac{\bar{p}_B(1 - \bar{p}_B)}{n_B}} \quad \text{(p. 120)}$$

12. Contingency tables, n rows, m columns:

$$\chi^2 = \sum_{i=1}^{m \cdot n} \frac{(o_i - e_i)^2}{e_i}, \quad \text{with } (m - 1)(n - 1) \text{ d.f.} \quad \text{(p. 154)}$$

13. ANOVA table:

Source of Variation	d.f.	Sum of Squares	Mean Square	F-ratio
Treatment	$k - 1$	SS(Tr)	$\dfrac{\text{SS(Tr)}}{k - 1}$	$F = \dfrac{(\quad)}{(\quad)}$
Error	$N - k$	SSE	$\dfrac{\text{SSE}}{N - k}$	

where k = Number of populations;
N = Total number of data;

$$\text{SS(Tr)} = \sum_{i=1}^{k} \frac{T_i^2}{n_i} - \frac{T^2}{N} \, ; \qquad \text{SST} = \sum_{i,j} x_{ij}^2 - \frac{T^2}{N} \, ;$$

$$T_i = \sum_{j=1}^{n_i} x_{ij}; \quad T = \sum_{i=1}^{k} T_i; \qquad \text{SSE} = \text{SST} - \text{SS(Tr)} \quad \text{(p. 208)}$$